MW00753677

Home Guide to

Plumbing,
Heating and
Air Conditioning

A Popular Science Book

Home Guide to

Plumbing,

Heating and
Air Conditioning

by George Daniels

POPULAR SCIENCE BOOKS

HARPER & ROW, PUBLISHERS

NEW YORK, SAN FRANCISCO, LONDON

ACKNOWLEDGMENTS

The author wishes to thank the following firms for their assistance in the preparation of this book: Sears, Roebuck and Co.; Montgomery Ward and Co.; American Standard, Plumbing and Heating Division; Crane Co.; The Donley Brothers Co.; Copper Development Association, Inc.; Grinnell Co., Inc.; The Heyman Co.; Ace Industrial Hardware Co.; American Gas Association, Inc.; Republic Steel Corp.; United States Steel Corp.; Committee of Steel Pipe Producers, American Iron and Steel Institute; National Oil Fuel Institute; Holman, Inc.; General Motors; Space Conditioning, Inc.; De Palma Brothers; Danbury Plumbing Supply Co.; Hoffman Fuel Co.; Westinghouse Electric Corp.; General Electric Corp.; Airtemp Division of Chrysler Corp.; Carrier Air Conditioning Co.; Philco Corp.; Fedders Corp.; Electromode; Dent Electric Co.; Stanley Tools; Plastics Pipe Institute; Lennox Industries, Inc.; Air Conditioning and Refrigeration Institute.

Copyright © 1967, 1976 by George Daniels
Published by Book Division, Times Mirror Magazines, Inc.

Brief quotations may be used in critical articles and reviews. For any other reproduction of the book, however, including electronic, mechanical, photocopying, recording or other means, written permission must be obtained from the publisher.

Library of Congress Catalog Card Number: 67-10841
ISBN: 0-06-010957-2

First Edition, 1967
  Nine Printings

Second Edition, Revised and Updated, 1976

Sixteenth Printing, 1983

Manufactured in the United States of America

Contents

Introduction

WE HAVEN'T had our modern plumbing, our automatic heating, and our air conditioning very long, and considering the obstacles that have impeded their development over the years, it's a wonder we have them at all. England, for example, once rated the burning of coal as a capital offense, and at least one Britisher was executed for the crime. So it's understandable that heating innovators may have become a bit timid. In France, advances in cooling rather than heating were blocked by the government. When 16th century Frenchmen built a thriving business by carting tons of snow and ice from the mountains to chill their summer foods and make frozen delicacies, they might soon have progressed to air conditioning by the same means. But the icemen's taxes were boosted until the business collapsed, and artificial cooling was forgotten for another hundred years. (More about the rigorous past of heating and cooling shortly.)

Plumbing, on the other hand, made slow headway largely because our ancestors of long ago rarely took much interest in it. If they had they might have copied an almost incredible example 4,000 years ago. At that time, archaeologists now know, the plumbing in the Cretan palace of King Minos was so advanced it incorporated the most important features found in our plumbing codes today. Its sewerage system, for instance, was vented as required by modern health regulations. Its toilets not only could be flushed, but, like all present-day plumbing fixtures, were designed to seal out sewer gases. And, for a final master's touch, the terra cotta water-supply pipes were tapered to increase the flow velocity at points where sediment might otherwise accumulate. But, along with the ancient culture that created them, all these things and their related ideas of sanitation, vanished. If they had not, many of the great plagues of history might never have occurred; and, in addition to the name of King Minos, history might have recorded the name of his plumber.

Though hygienic refinements faltered and progress was slow, the water-supply aspects of plumbing eventually revived on a major scale. What was probably the first long-distance municipal water system was developed for Jerusalem by King Hezekiah in 727 B.C. To do it, he had his workmen tunnel a third of a mile through rocky hills to bring water from the Pool of Siloam to the city. The finished job worked beautifully.

To the ancient Romans, however, goes the credit for first supplying water in the truly grand manner. They not only did their plumbing on an unprecedented scale, but coined the word itself, and developed what was undoubtedly the first method of cheating the water company. Around Nero's heyday (34 to 68 A.D.) they were getting their water through as many as fourteen aqueducts with a total length of almost 360 miles. Siphon systems lifted the flow over hills, towering stone trestlework carried it across valleys. Approximately 130 million

1

gallons a day poured into the city where more than 90 million gallons went into 247 reservoirs, nearly 40 decorative fountains, and close to 600 public water-supply basins operating around the clock. At the basins, official tenders supplied the Roman man in the street with his water on a fill-your-jug-for-cash basis. And, as the business brought in a total close to $40,000 a year, shady characters occasionally found their way into it. As might be expected, they also found a way of boosting their profits by "beating the meter." The "meter" in those days was simply a piece of pipe a little less than a foot long out of which water flowed constantly from the main. As the meter pipe was made of soft lead not quite an inch in diameter, the slick water tenders simply enlarged it with a little prying on nights when things were slow. And from then on, they enjoyed a greater supply of water and a more rapid procession of customers than the city fathers had planned. So Rome's water officials had to switch from lead pipe to harder brass and bronze.

But it was lead, known to the Romans as *plumbum*, that gave plumbing its name. The ancient Roman plumber, however, was called a *plumbarius*, and frequently was a woman. At least, we know that many Romans of the fair sex owned plumbing shops, as much of the pipe unearthed from the ruins bears the stamp of a feminine name. Their best customers, of course, were the government and the wealthy citizens who had water piped directly to their homes. Along with their water supply, the latter received a status symbol of sorts by having their names stamped into the pipe together with that of the plumber. As the pipe was exposed, the monogram also made it awkward for less-than-honest Romans to tap into it for their own benefit. And it seems that Romans with such inclinations were rather plentiful. In fact, Sextus Julius Frontinus, one of Rome's most dedicated water commissioners toward the end of the century, found some of them bleeding the aqueducts before they even reached the city, and others tapping into the city mains and fountain supply pipes — while some of his own hirelings were juggling the metering pipes at distribution points for a tidy profit. In any event, the Roman plumbers were efficient. They made their own pipe by rolling a lead sheet around a wooden cylinder and soldering the seam. They did it so well that their ancient pipes, tested in modern times, have withstood pressures as high as 250 pounds per square inch — enough to blow your car's tire to shreds.

Much of Rome's prodigious water supply went into the baths for which the city's inhabitants were famous. There were, however, no bathtubs in the luxury homes of the day as it was customary to merely flood the bathroom and use it as a pool. A roaring fire under the concrete floor warmed the water, and the excess flame and smoke rose through the hollow tiles behind the marble walls and created a sort of sauna. But this was a family affair. Important guests were taken out to the lavish public baths that served as the men's clubs of the era. Some of these, whose pools were replenished through bronze pipes with silver spouts, could accommodate better than 3,000 bathers in a building about the size of New York's old Penn Station — which, incidentally, embodied many architectural features of the Roman bath clubs. Some of the luxurious bath establishments reserved certain hours for women — and a few operated on a coed basis.

The bath and its plumbing, however, all but disappeared with the Roman Empire, condemned by many Christian churchmen as immoral. The bathtub, apparently first used by the Greek athletes of ancient Nemea, became an unmentionable thing through Europe's Dark Ages. But, with filth and pestilence rampant, there were many wise defenders of the tub. England's Henry IV struck a blow in its favor in 1399, when he created the military order of the Knights of the Bath. The selected warrior was assured at least of a shave, a haircut, and a good tub bath, as all three were part of the ceremony that pledged him to purity when he became a member of the order.

Centuries passed before the bathtub could shed its evil stigma. Even in the 19th century, long after aeronauts had ridden horses into the sky slung under balloons, crossed the English Channel by air, and set an altitude record of nearly five miles, you still needed a doctor's prescription to use a bathtub in Boston. So we can appreciate the courage of Britain's Sir John Harington when, almost three centuries earlier, he sought acceptance unsuccessfully, for his invention – the water closet. But, unlike many others who helped make our modern plumbing possible his name is at least recorded, and his contribution has made the grade.

While all this was going on, the science of heating was muddling through. Because man urgently needs to be warm, it got its start long before plumbing – apparently in the Stone Age. Evidence indicates that our relatives of that period made use of fire, though at first they probably had to rely on Nature to create it by lightning. Later, we know they brought fire into their cave homes, for both light and heat. And their relics tell us that their combination lamp and heater was often a skull stuffed with fat-soaked moss. The skulls, charred from use, have been found on cavern ledges, and evidently provided warmth and illumination for the primitive cave-wall decorators whose art work has been discovered in modern times.

Wood has usually served as fuel for the major heating jobs of the past, but just about everything else that burns has also been used, including the dried manure that has fed the shepherds' fires in treeless lands over the centuries. Even in a primitive shelter, the problem of the warming fire was smoke, for the true chimney wasn't invented until the 1300's and three more centuries passed before it was available to the general public. In the meantime, plain folks in most of the civilized world led either a cold or sooty existence. The pit dwellers in parts of Europe and Russia, residing in crudely roofed holes in the earth, warmed themselves with a fire on the mud floor. When their heating system threatened them with suffocation they simply banged a hole in the roof.

Elsewhere, a permanent hole in the center of the roof, with a hearth directly below it, improved things somewhat. But while the smoke escaped through the hole most of the heat went with it and the rain came in. So the hard-put householders capped the hole with a little open-sided steeple called a "smoke louver" to let the smoke out without letting the rain in.

The chimneyed fireplace, when it finally appeared, was definitely better than a hole in the roof. But it, too, had its troubles – and its detractors. There were many who claimed, for example, that without the usual cloud of smoke at the ceiling to preserve the rafters like smoked hams, the roof would rot and cave in. And in England the rip-roaring log burners depleted the national firewood supply so rapidly that the government urged people to burn coal—instead of executing them for doing it.

In Paris, Dr. Louis Savot brought modern heating a step closer in the early 1600's when he installed the first heat circulating fireplace in the Louvre Palace, using ducts as we use them today. (Modern types like the Donley Heatsaver and the Heatilator work the same way.) Around the same time, another Frenchman, Gauger, designed an eliptical fireplace to beam heat into the room like a parabolic reflector. It also drew part of its air supply through a duct from out of doors to combine heating with fresh air circulation. Although neither of these features survived, the translation of Gauger's writings introduced a new word—ventilation.

Probably the greatest combination of improvements came from colonial America's Benjamin Thompson. Born poor, but with many talents, Mr. Thompson correctly calculated a lunar eclipse at 13, headed New Hampshire's state militia in his early twenties, became Britain's undersecretary of state a few years later, and was knighted by the king. To round out his career he became Bavaria's minister of war, minister of police, grand chamberlain, and a count of the Holy Roman Empire. In his less pressing moments he founded England's Royal Institution, Harvard's Rumford professorship, and married the wealthy widow of France's renowned but guillotined chemist, Lavoisier. Somehow, he also found time to make a careful study of more than 500 smoky London chimneys and recommend improvements. And although his personal career may be a bit hard to follow, his heating innovations are not. He greatly reduced chimney heat losses by narrowing the throat where the top of the fireplace joins the flue. He splayed the fireplace sides to produce Gauger's heat reflective effect with far simpler brickwork. And he sold the public on the idea of matching the fireplace to the room size. All of these ideas worked so well they guide the construction of practically every fireplace to this day.

To another Benjamin, Benjamin Franklin, goes the credit for a major transition in heating—from the fireplace to the far more efficient metal stove. Standing free of the wall, it radiated heat in all directions and soon became a familiar fixture in American Colonial households. Many a parlor stove warmed the upstairs bedrooms, through an iron grille in the second floor. Just who first thought of putting the stove in the cellar with a metal shell around it to funnel heat to the house above, isn't known, but he was the originator of central heating.

Long after we acquired our automobiles and our airplanes, we were still shoveling coal to keep warm. Perhaps the most important step toward automatic heat took place in 1861. This was the invention of the oil burner—by a German mechanic named Werner, employed in a Russian oil refinery. Its advantages, including the absence of ashes, prompted industry to welcome it with open arms. But it still required a trained attendant, so the homeowner labored on with his coal shovel and kindling for almost forty more years. The fully automatic oil burner finally made news in the same year another historic event hit the headlines, 1919—when Alcock and Brown made the first trans-Atlantic airplane flight.

Gas heat was still in the future as most of the gas used for cooking was then of the manufactured variety, and quite a bit costlier than the natural type. But natural gas had entered the picture. America's first natural gas well, in fact, was apparently started in 1821 by an Edward Howard, of Fredonia, New York, who was looking for a better water supply. But streams of bubbles rose from the water in Mr. Howard's well and burned merrily when lighted. Understandably perplexed, Howard called in a local gunsmith named William Aaron Hart, who decided to dig deeper in the hope of finding more gas. And he did. At a depth of 17 feet he found so much gas it actually hissed. At 27 feet he found so much he quit and capped the well.

Fredonia's first gas pipes were hollow logs leading from the well to nearby buildings where it was burned experimentally. Although not tried for heating, it proved to be a useful lighting fuel, so Hart replaced the logs with lead pipes and extended his supply lines to other buildings in the town, including the Abell House, Fredonia's inn. Compared to the candles and whale oil lamps of the day, the gas light was dazzling. In fact, when the Marquis de Lafayette visited Fredonia in 1825 and stayed at the Abell House, he was fascinated by the lighting, and inquired as to the source of the lighting fuel. Told that it came from a mysterious source under ground, the Marquis remarked with a chuckle that perhaps he'd better leave the town posthaste, as it was evidently connected to hell.

Home air conditioning didn't make its appearance until 1929, ten years after automatic heating. The pioneer was a nifty Frigidaire model that tipped the scales at about 640 pounds and sold for around $600. It cooled the room—the way you can do it today with a window model you can pick up and carry.

As with home heating, the caveman seems to have been first to enjoy air conditioning. The cool, even temperature of caves has been used by cheese and wine makers over the centuries in aging their products. But true, controllable air conditioning was a long time coming. Its grand father, the ice-making machine, might have been a hundred years late in developing if it hadn't been for the Civil War and a freak American winter. Many famous experimenters, including Michael Faraday and James Watt, had discovered ways to create low temperature artificially. In 1834, an American born engineer, working in London, patented a very practical ice-making machine that quickly found a market in breweries and meat-packing houses. But the public, with only a vague knowledge that ammonia was somehow used in the ice machine of the day, refused to adopt it. Instead, they bought the natural winter ice then being shipped, packed in sawdust, by Clipper ships to every major port in the world.

When the Civil War broke out, however, the South could no longer get its ice from the North. They managed to slip through the naval blockade with one of the first absorption-type ice machines (like a giant gas refrigerator) built in France by Ferdinand Carré, its creator. With no natural ice available, southerners soon learned that artificial ice, as it was called, was just as good and often better, because it was cleaner. Those who still shunned it were finally convinced in 1890, when the North had a winter so strangely warm that almost no natural ice formed, and a frightening chain reaction began. The window boxes and outdoor pantries that usually preserved foods through the winter were no longer cold, and their contents spoiled. Then the tainted foods caused illness and fever of almost epidemic proportions. The natural ice commonly used to make fever patients more comfortable was not to be had, so the doctors turned to artificial ice—available at the time from industrial users—and we've been using it ever since.

1 | How Your Plumbing Works

THERE's nothing complex about household plumbing, but it's easy to misunderstand or overlook some of its important features — unless you know how to spot them and how they work. After that you're likely to find it much simpler to diagnose and cure common plumbing troubles, and you won't need to be timid about tackling improvements or additions to the system.

THE WATER-SUPPLY SYSTEM. As your plumbing really begins with the water supply, we'll start there. The pressure that drives the water through your pipes to your plumbing fixtures is relatively high — usually higher than the pressure in your car's tires. In private well-pump systems pressure comes from air compressed above the water in the supply tank. The pump bubbles in air along with the water, automatically maintaining the right amount in the top of the tank. If the water comes from a municipal water supply, gravity usually provides the pressure, through high-land reservoirs or tall water towers kept filled by pumps. Whatever the pressure system, the result is the same at your faucets — driving power that averages around 40 pounds per square inch. That's enough to shoot a sizeable stream across your lawn or over the roof of the house.

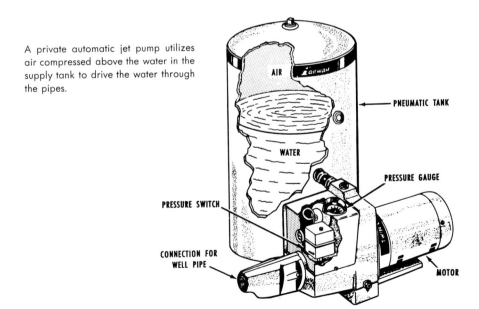

A private automatic jet pump utilizes air compressed above the water in the supply tank to drive the water through the pipes.

AIR

PNEUMATIC TANK

WATER

PRESSURE GAUGE

PRESSURE SWITCH

CONNECTION FOR
WELL PIPE

MOTOR

7

The volume of water per minute — the time it takes to fill a teakettle or a tub — depends also on the length and diameter of your water pipes and the height of the outlet from which it flows. And since this fast-moving water weighs more, volume for volume, than the heaviest timber in the house, it carries more force than most of us realize.

A 40-pound pressure can drive almost 3½ cubic feet of water per minute through a typical ¾″ household water pipe. You can draw your bath in about a minute and a half, but you have more than 200 pounds of water streaking through the pipe every 60 seconds. If you shut it off abruptly, it stops with a bang that actually shakes the pipes and resounds all over the house. Plumbers call the bang "water hammer" and eliminate it with one of the plumbing devices most of us rarely see, and probably wouldn't recognize if we did. The device is an "air chamber," a capped branch of pipe a little over a foot long, extending upward from the regular water pipe, close to the faucet. As it contains only air when it's capped, and as the "entrained" air in the water keeps it filled with air, water can get into it only by compressing the air. And this compression provides a cushion that eliminates the bang when you turn off a faucet suddenly. Unless you know what these air chambers are, however, you might think they were just evidence of false starts made by a plumber who changed his mind about extending the water pipes.

Emergency shut down. To stop the flow of hot or cold water out of a leaky faucet or burst pipe, close the main valve (in the water-service line), then open one or more hot- and cold-water faucets below the level of the leak and a hot and cold faucet on the top floor. Closing the main valve stops the flow of water into the system. Opening the faucets above and below the leak drains the pipes in between. Should the leak be on the top floor with no pipes above, just closing the valve in the pipe leading to the leak, or just closing the main valve, will stop the flow of water. Should the main valve itself become defective, or should there

Take the trouble to locate and try the main valve before an emergency arises. Should a pipe burst, you will need to close the valve in a hurry.

be a leak in the associate piping, phone the water company and have them shut off the water at the curb valve.

Emptying the system. To drain the hot- and cold-water system prior to leaving the building unheated through the winter, first turn off the water at the main supply valve. Then flush all the toilets, open all the faucets, and open the drain valves. The open faucets admit air to the fixture supply pipes, and the flushed toilets do the same by opening the flush tank inlet valves. Otherwise water would remain trapped in the small supply pipes to freeze and break them later. (A soda straw can show you how this works. Immerse the straw in a glass of water, then hold your finger on top of the straw to seal it. No water leaves the sealed straw when you lift it out of the glass because no air can get in. A partial vacuum holds the water. But lift your finger from the top of the straw to admit air, and the water empties immediately. Many a home water pipe has frozen and burst because this principle was neglected.)

Preventing trap freeze-ups. If the house must be left unheated at any time during winter weather, the traps, like the water-supply system, must be protected against freezing. But draining them would admit sewer gas. So automotive permanent-type antifreeze does the trick. To be sure of safe water–antifreeze proportions, it's best to drain the traps first. If they have drain plugs this is easy. If not, fill a small-diameter rubber hose with water, and slip one end down into the trap while pinching the other end closed. Place the pinched end in a bucket below the trap, release it, and water will siphon from the trap to the bucket. Then you can pour enough pre-mixed antifreeze into the trap to fill it. If you have a small boat bilge pump you can use it to empty the toilet bowl. Then pour in the pre-mixed antifreeze. The flush tank, of course, should be emptied and sponged dry before this by flushing it after the water has been turned off.

The hot-water system is merely a branch of the cold-water line leading through a water heater. One of the earliest types provides a good illustration of the basic principles. This consisted simply of a hot-water storage tank with a cold-water supply pipe leading into it near the bottom, and a hot-water pipe leading out of it at the top. The tank was heated like a giant teakettle, by a gas burner underneath it. Heated water from the bottom rose by natural convection to the top, from which point it was piped to the hot-water faucets. Later, the arrangement was improved by connecting a vertical coil of copper tubing to the side of the tank, leaving the cold- and hot-water connections the same as before. The flame, however, was placed under the coil. Cold water from the tank entered the bottom of the coil, acquired heat from the flame, and re-entered the tank from the top of the coil. As the coil conducted the heat to the water more rapidly than the earlier system, it provided more hot water per dollar's worth of fuel. And, as an enclosure could be built around the coil, and connected to a chimney, any type of fuel could be used. Heaters of this type are still widely used. The modern "package" type of water heater, though varied in design, works on the same basic idea. Electric heaters, however, get their heat from an electrical resistance unit that projects into the tank itself. This operates like the small electric immersion heaters many barbers use to heat a tumbler of water for shaving. There are also "tankless" heaters that operate automatically whenever the water flows.

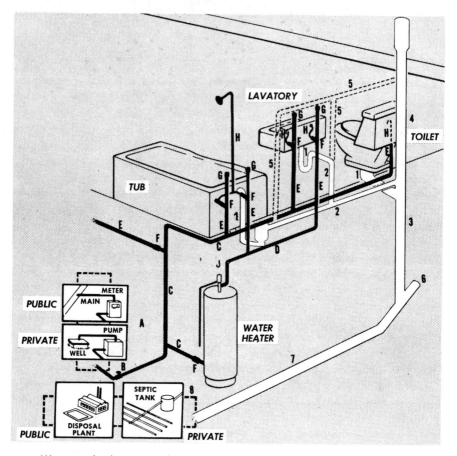

Water supply, drainage, and vent pipes for plumbing system in a one-story house.

Water-Supply System

A. Water-supply pipe.

B. Stop and waste valve to shut off water and drain system. It must be at lowest point in system.

C. Cold-water main to fixture.

D. Hot-water main to fixture.

E. Branch lines from mains to fixtures.

F. Shut-off valves where likely to be needed in case of repair.

G. Air chamber to eliminate water hammer when water is shut off suddenly at faucet.

H. Fixture supply line adapted to each specific fixture.

J. Relief valve.

Drainage System

1. Fixture drain matched to fixture and including trap, unless fixture has built-in trap.

2. Branch drain pipe from fixture to soil stack.

3. Main soil stack, into which all branch drains flow.

4. Main vent in upper portion of soil stack through which air flows for free drainage and from which sewer gas escapes.

5. Re-vent pipe that draws air from main vent for easy drainage from fixtures.

6. Cleanout plug — always required at base of stack.

7. Building drain carries waste to disposal system or sewer.

8. Final disposal may be through public sewer or septic system.

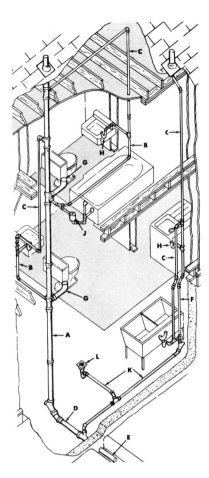

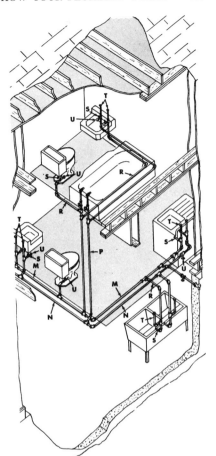

Parts of a typical drainage system in a two-story house with basement.

Parts of a typical water-supply system in a two-story house with basement.

A. Soil stack	F. Waste stack
B. Waste pipes	G. Closet bend
C. Vent pipes	H. "P" trap
D. House drain	J. Drum trap
E. House sewer	K. Branch drain
	L. Floor drain

M. Cold-water main	R. Fixture branches
N. Hot-water main	S. Fixture supply pipes
P. Risers	T. Air chambers
	U. Shut-off valves

Another water-heating device is widely used in homes whose winter heat is provided by hydronic (forced hot water) heating systems. In these, a coil of tubing mounted inside the main heating boiler is connected to a hot-water storage tank on the outside. Heat is transferred through the coil from the boiler to the hot-water supply for the faucets. But there is no mixing of boiler water and faucet supply water. Shutting off the pump that drives boiler water to the heating system radiators enables the unit to heat the faucet supply in summer without heating the house. The system can also be supplemented by a regular water heater if unusually large amounts of hot water are required. When this is necessary, the important factor is "recovery" rate—how long it takes the unit to provide a new supply after all hot water has been used. This depends on the fuel. Oil does the job fastest, gas ranks next, and electricity last.

A "relief valve" should be connected to the hot-water storage tank even on automatic units. (If it's absent it should be added.) Although some of these valves operate by pressure only, extra protection can be provided by those that respond to either pressure or temperature. Both types are small, about the size of a banana. The valve's purpose: to release water to a drain whenever the temperature approaches boiling or the pressure exceeds the normal level. This prevents burst tanks and other damage to the system. When the temperature and pressure return to normal the valve closes.

A hand-operated drain valve at the lowest point of the heater empties it if the house is to be left unheated. All hot-water faucets should be opened during the process to clear the supply pipes. When the water system is re-activated, it is extremely important that the entire water heating system be filled with water before its heat source is turned on. You can tell when it's full as water will run from the hot-water faucets, which should be left open during the filling process.

THE DRAINAGE SYSTEM is the part of your plumbing that carries off the waste from all your plumbing fixtures. Its main artery is the "soil stack." This is a large-diameter pipe (usually 4") that starts from a basement connection with the house sewer and runs vertically upward through the roof to the open air. The upper end, called the "main vent," is open. All plumbing fixtures, including toilets, located near the soil stack drain into it. Fixtures like sinks and washtubs, located at a distance from the main stack, may drain into a smaller vertical pipe called a "waste stack." This, like the main one, runs from a basement sewer connection upward through the roof and is left open at the top. These stacks are

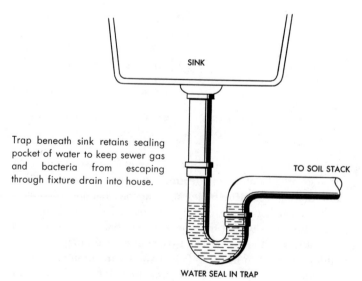

SINK

Trap beneath sink retains sealing pocket of water to keep sewer gas and bacteria from escaping through fixture drain into house.

TO SOIL STACK

WATER SEAL IN TRAP

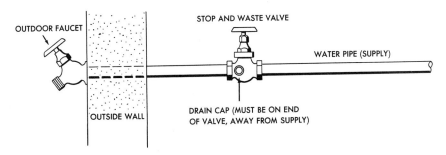

When the supply pipe for an outdoor faucet must run under an outside, unheated porch or through an unheated crawl space, an ordinary sill faucet is used. This mounts on the sill of the house framework and has a stop and waste valve inside the basement. Faucet must be shut off and the water drained from the supply pipe before freezing weather begins.

left open at the top for several reasons. First, they admit air so the downflow of drainage isn't retarded by a partial vacuum above it. (As in the soda straw example mentioned earlier.) Second, they prevent a build-up of sewer-gas pressure in the drainage system that might force its way into the house through the fixtures. (The gas flows out through the open vents.) And third, by admitting air freely at their upper ends, the stacks prevent the drainage downflow from sucking air through the drain pipes from sinks, basins, and tubs, and sucking water out of the "traps" beneath them.

What the traps do. The traps in your drainage plumbing are the U-shaped sections of pipe under your sinks and basins. (There's one under your bathtub, too, but you can't see it.) They act as automatic seals to keep sewer gases from drifting up through drains and into the house. As water always remains in the bottom of the U it closes the pipe so the gas can't pass through. (The water that remains in a toilet bowl does the same job.) If the system is inadequately vented, however, the downflowing waste sucks additional air after it through the traps.

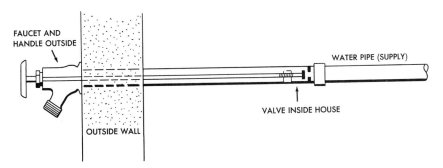

When the faucet is located directly outside the basement wall, a freezeproof faucet may be used. This has the faucet handle outside with the faucet shaft extending to the actual valve inside. When you turn it off, water is shut off inside basement — where it's warm — and all water remaining in the faucet body drains out, preventing freezing. Thus the faucet may be left open all winter.

If you hear a glugging sound immediately after a fixture empties, that's what is happening. This may leave too little water in the trap to seal out the gas, and a health menace results. (If you have a glugging fixture in your house always run about a cup of water into it after the glugging stops. This reseals the trap without restarting the glugging.)

To make certain that fixture traps are not sucked dry by outflowing waste running through their own drain pipes to the main stack, individual fixture vents are often used, especially if the fixture is more than a few feet from the stack. The fixture vent connects into the fixture drain very close to the trap and runs upward inside the wall to connect into the main stack somewhere above the fixture.

THE OUTDOOR FAUCET for the garden hose is one more essential feature of most household plumbing systems. This is often called a "sill cock" as it is mounted on the wooden sill of the house structure. In cold areas a special type of additional valve is provided in the sill-cock supply pipe inside the house. This is called a "stop and waste" valve. It shuts off the water to the outside faucet and also provides a vent to admit air to the intervening section of pipe to drain it and the outside faucet completely. The vent is opened by removing a small brass cap on the side of the valve. (If you install this type of valve *be sure* to mount it so that the vent is on the *outgoing* end of the valve.)

If you're a new homeowner it's wise to check over your plumbing system as thoroughly as you can with the foregoing pages as a guide. More than one house-holder has found an important detail missing—a drain valve, for example. In many homes, too, the bathtub trap is completely inaccessible. Often it can be made accessible through a small removable panel in the wall of an adjacent room or closet. It doesn't take long to correct minor oversights like these. But it seems a lot longer if you have to do it in an emergency, or on the day you planned to leave for that Florida vacation.

2 | Plumbing Tools

IF YOU have any household fix-it tools, the chances are you already have a few you can use on common plumbing-repair jobs. A screwdriver is one. And an adjustable wrench big enough to fit the hexagonal caps through which your faucet shafts turn is another. You can buy a plumber's force cup to clear clogged drains, and a plumber's flexible metal "snake" for the stubborn drains the force cup won't clear, for the price of a couple of neckties. These, plus a pair of ordinary pliers, will enable you to meet the usual plumbing emergency with a good chance of success. For example, you can fix a leaky faucet, stop a singing toilet (in most cases), and drain an overflowing sink. Beyond this, your best bet is to buy your plumbing tools as you need them, and rent the expensive ones you're not likely to use again. You'll rarely need an expensive tool, however, unless you're actually replacing a major part of your plumbing system or adding something to it.

Just which plumbing tools you buy or rent depends on how your plumbing was originally put together. If your pipes are copper with solder-sealed fittings, you'll need a propane gas *torch* to solder the connections in any changes or additions. If you plan to assemble any cast-iron soil pipe, as you might if you add a lavatory, you'll need a plumber's furnace and a little melting pot for the lead that seals the joints. And you'll also need a caulking iron and a hammer to spread and tighten the lead in the joint after it cools. The torch is handy for many jobs other than plumbing — paint removing, for instance. And the caulking iron costs less than half the price of a carton of cigarettes.

If your pipe has threaded connections, like galvanized steel, you'll need a pair of stillson wrenches. These have pivot-jointed jaws that tighten their grip on round objects like pipe when the pull on the wrench handle is increased. You need one to hold the pipe, another to grip the fitting. For large-diameter drain pipes you may need a pair of "chain" wrenches. These get their grip with a toothed chain that wraps around the pipe and draws tight when the handle is pulled. Measure the pipe diameter you'll be working on, and buy or rent your wrenches to fit it. You'll also need a pipe vise to hold the pipe, a pipe cutter, and a pipe threader to thread the ends for the fittings after you do the cutting. As these are among the more expensive tools, you'll usually save money by renting them unless you are installing a complete plumbing system.

Whatever type of pipe you're using, you'll need a pipe reamer. This is a conical tool with cutting blades converging at its point. Mount it in the chuck of an ordinary carpenter's bit brace, and turn it inside the cut end of the pipe to remove the rough burrs formed by the cut. You might as well buy this tool as it comes in handy on other jobs — such as enlarging a hole you've made in wood or sheet metal when your biggest drill bit was undersized.

15

Pipe cleaning tools: force cup for minor stoppages (left); "snake" for stubborn stoppages.

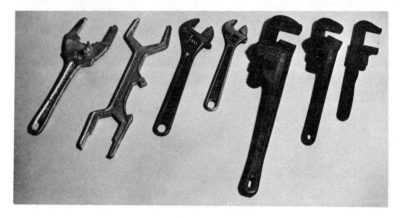

Wrenches commonly used on plumbing jobs (left to right): adjustable spud wrench (for all very large hex-shaped parts); solid spud wrench, matched to standard spud sizes only; adjustable wrenches (two) for gripping smaller hex fittings like flare fittings, faucet caps, packing nuts; Stillsons (two shown, large and small) for gripping pipe and round fittings, not for hex shapes; monkey wrench, for hex fittings and other jobs.

Another very useful plumbing tool is an old-fashioned monkey wrench. These are getting hard to find today except in the high-grade (and high-priced) versions made specifically for plumbers. They have a major plumbing advantage, however, in that their jaws open much wider than those of the fancy, modern adjustable wrenches of comparable size. This lets them grip the many hexagonal plumbing fittings too big for anything else but a special wrench. Some of these are too big even for the monkey wrench. You may find such a fitting on the "spud" pipe that joins your toilet flush tank to the bowl. If so, you can buy a solid or adjustable "spud wrench," which isn't too expensive.

From this point on, as with the simple repair tools, it pays to acquire special items only when you need them. Some plumbing tools you may never need. And, if you happen to be buying your plumbing materials from a mail-order house, look into any tool plan they may have. Some of them lend you the tools you need to do major plumbing jobs with materials you buy from them. One advantage to getting your tools in this way (in addition to the economy) or from a tool-rental company, is that you can also get specific instructions on how to use them. In case you buy the tools, however, the following pages will guide you in using them.

STILLSON WRENCH. To those who have never used it before, this tool seems like a loose, wobbly monkey wrench, but it gets one of the most powerful grips possible on a pipe. There's a simple trick to adjusting it, however, that's not often explained. Set the jaws so the pipe jams between them about halfway back. Then, when you pull on the handle the jaws tighten together, as intended, and bite into the surface of the pipe. It's easily possible to get the jaws too tight or too loose for their maximum grip, but a few trials will give you the knack. As these wrenches really "bite," they leave their teeth marks on the pipe, but there's no harm done. You're not supposed to use a stillson on hexagonal fittings or on nuts or bolts because the jaws may skid over the corners of the hex and round them somewhat as they tighten. But if you're faced with an otherwise impossible job go ahead with your stillson. It can get a hold on almost anything. And if no other wrench will fit it afterwards, you can always use the stillson again. It's a tool you'll get to like.

GAS PLIERS. Gas pliers differ from dozens of other kinds in having serrated, curved jaws to hold pipe. Standard gas pliers are handy for working with small nuts and bolts and loosened pipe. A far more useful type is a patented design called Channel-lock, which have widely adjustable jaws offset almost 90 degrees from the handle. Although offset adjustable gas pliers are not unusual, Channel-lock pliers differ in that the adjustment is positive; the jaws will not slip apart no matter how hard you squeeze the handles. This makes it possible to secure almost as good a grip on a pipe with these pliers as with a stillson wrench.

To vary plier opening the handles must be spread to their full extent. With the handles spread wide, the jaw opening can be varied by sliding one handle over the other.

Ten-inch pliers are the most useful size for plumbing. They open to 1½

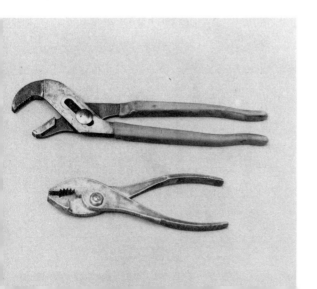

Types of gas pliers. The pair at top are Channellock pliers, which will not open beyond their setting no matter how much pressure you apply. Standard gas pliers below cost less but are not as strong.

Chain wrench for gripping pipe too big for the average Stillson wrench. Chain is pulled snugly around pipe and hooked in end of wrench handle. Rocking action of handle draws chain tighter, presses serrations in end of handle into surface of pipe for nonslip grip.

inches. You will probably find this tool the one tool in your kit that you use most, after the screwdriver.

CHAIN WRENCH. These vary somewhat with the manufacturer, but the basic procedure is the same. Be sure to read the directions that come with your wrench. With a typical model, set the toothed crotch on the business end of the handle against the surface of the pipe. Then pull the attached chain around the pipe as snugly as possible, and hook it to the nib on the handle tip. It may not seem very snug at this stage, but when you pull on the handle the crotch rocks up on a high point in such a way as to tighten the chain like a vise. Then those teeth really bite in. The chain wrench can usually grip a pipe two or three times the diameter possible with a stillson of comparable handle length. So it's your best bet for big drain pipes.

PROPANE TORCH. Read the manufacturer's instructions, as different makes vary somewhat.

Usually the procedure is as follows: Screw the torch burner snugly on to the gas cylinder fitting. Then open the control valve just enough to produce a low hiss of escaping gas and light it immediately. Let the burner heat up fully

Propane torch consists of two parts: the metal bottle that holds the gas and is replaceable, and the valve and nozzle that is screwed onto the bottle. For safety, always store the torch disassembled.

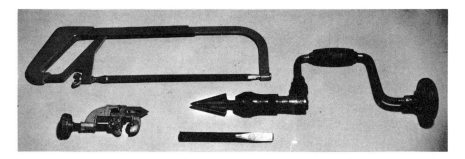

Cutting and reaming tools: upper left, hacksaw; lower left, tubing cutter; upper right, bit brace with pipe reamer bit — for smoothing inside of pipe end after cutting; lower left, cold chisel for cutting cast-iron pipe after hacksawing part way through.

before you open the valve to larger flame size. Disconnect the burner from the gas cylinder when you're not using it, and don't store the tank in the living area of the house. (Complete details on making leak-proof pipe connections with a propane torch are given in Chapter 3.)

PIPE REAMER. First, provide a firm means of holding the pipe. A pipe vise is best. But you can also buy removable pipe jaws for many ordinary workshop vises. Mount the reamer shank securely in the bit brace chuck. Insert the point of the reamer into the end of the pipe as far as it will go. Try to keep its axis parallel to the pipe as you turn it *clockwise* with moderate pressure. And stop reaming as soon as the burr has been removed from the inside of the cut pipe end. If the pipe is to be threaded, the reaming job should be done *after* threading.

PIPE CUTTER. This tool is far superior to a hacksaw for pipe cutting because it assures a right-angle cut — very important for threading and most other connection methods. The cutter head contains a pair of small rollers that bear against the surface of the pipe, and a very sharp and hard steel cutting wheel. Place the cutter on the pipe with the cutting wheel exactly on the measured cutting line. Then tighten the cutter's screw handle just enough to force the cutting wheel slightly into the pipe surface. Give the cutter one full turn around the pipe at this setting, establishing a shallow cut. Tighten the cutting wheel a little deeper into the pipe before each successive turn until the pipe is cut through. Apply *thread-cutting oil* to the pipe and the cutter at intervals during the operation. This eases the job and lengthens cutter life. You can buy cutting oil where you buy your pipe. If you're working with copper pipe or tubing, cut it in the same manner, using a "tube cutter." This is similar to, but lighter than, the pipe cutter.

PIPE THREADER. Most often called a stock and die, this tool is used to cut male pipe thread on galvanized, black, steel, brass, copper and (occasionally) plastic pipe. The pipe is held in a pipe vise. A machinist's vise can also be used if the pipe size is such that the pipe can be jammed beneath the vise jaws. If

1. Clamp the pipe in a vise and remove the internal burrs with a pipe reamer, as shown, or a rattail file.

2. To start the die, place the larger opening over the pipe end and apply pressure with one hand as you turn with the other.

3. Always lubricate the die generously. If you cut the thread dry the thread will be ragged, uneven and may even leak. Die life will be greatly shortened.

4. Run the thread up until you can see two threads beyond the die itself or the end of the pipe is flush with the outside of the die.

not, a wooden jig can be used to clamp the pipe firmly between the flat, parallel vise jaws. The pipe end to be threaded must be square and free of internal and external burrs. Both the hacksaw and the pipe cutter produce internal burrs, which if not removed greatly restrict the flow of water. The hacksaw does not produce external burrs, and neither does the pipe cutter if moderate pressure is used. Use a rattail file or a pipe reamer to remove the internal burrs. Use a flat file on the external burrs. If these are not removed you may not be able to properly position the die.

A die and guide are selected to match the size of the pipe that is to be threaded. The die and guide are secured in the stock. Usually, all you need do is to loosen a thumbscrew to remove the old die and guide and tighten the screw after you have installed the proper size die and guide.

Position the guide and the die on the end of the pipe. Give the die and pipe end a squirt of cutting oil. If your plumbing supply shop doesn't have any, use a mixture of motor oil and kerosene, or a high-detergent motor oil. Never cut the thread dry or you will wear out the die and produce rough, ragged threads that will probably leak after you have assembled the joint.

Now press the die against the end of the pipe and give it a half turn to the right (clockwise when facing the pipe end). Then give the stock and die a quarter turn backwards, followed by a half turn forward. The back and forth progression is necessary to clear the die of metal chips. Lubricate the die and pipe every few turns. Continue the forward-backward-lubricate procedure until about five full-depth threads have been cut. It is not important to have exactly five; one more or one less does not matter. But if you cut too many you will not be able to tighten the pipe in the fitting. If you cut too few, the joint will be weak and may leak. With some dies the correct number of threads are indicated when the pipe end is flush with the surface of the die. Others require the pipe end to project two threads beyond the die's surface.

One way to determine whether or not you have cut the correct number of threads is to apply pipe dope (joint compound) to the new thread and run a female coupling or fitting onto the pipe as far as it will go with light wrench pressure. Then remove the fitting and count the threads that entered. If four

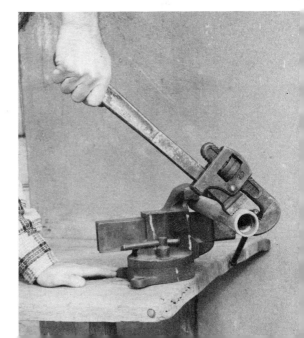

To assemble a screw-thread joint, apply pipe dope to the male threads and screw the female fitting in place. Use a Stillson wrench to tighten the joint, but do not over-tighten. Twenty or so pounds on the end of an 18-inch wrench on 1-inch pipe is plenty.

threads entered, you have cut the correct number of threads. Another method is to match your freshly cut threads against a commercially threaded piece of pipe of the same size. If you haven't cut sufficient thread, you can run your die back up and cut some more. If you have cut far too many, you can hacksaw off the excess. Then use a pipe reamer to remove the internal burrs and a flat file to clean up the external thread where the saw may have damaged them.

TUBE BENDER. Soft copper tubing can easily be bent with your hands alone. However, doing so will almost invariably flatten or kink the tubing, which of course renders it just about useless. To prevent the tube from flattening, use a tube bender. This tool consists of a long, tightly wound steel spring. It is slipped over the tubing at the point you wish to bend. Then the spring with the tube inside is simply bent with hand pressure. Afterwards, the spring is removed. Turning the spring facilitates its removal.

The only precaution you need observe in using the tube bender is that it isn't too large in diameter for the tube to be bent. If there is too much space between the tube and the spring, the tube will flatten when you bend it. Tube benders are manufactured in a number of diameters, so there is no problem in selecting the correct size.

To hand bend rigid copper tubing, first heat the section to be bent with a propane torch. This will remove the temper and you will have no difficulty bending the annealed tubing.

All you need do to use a spring-type tube bender is to slip the tubing inside the bend. If you cannot easily remove the tubing after bending, try turning the spring.

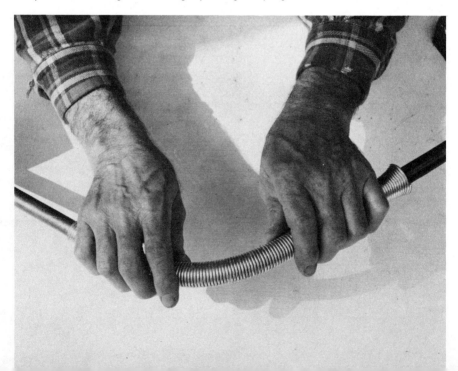

3 | Plumbing Materials and their Use

AT LEAST eight different kinds of pipe (plus variations) and as many methods of connection are used in modern residential plumbing. Some serve specific purposes, others compete with each other, and here and there, some are banned by local plumbing codes. If the picture seems baffling, it really isn't. The type of pipe a builder uses may depend merely on his budget. Or it may depend on some strictly local factor in the soil or water that gives one material an advantage over another. Usually, however, you have a wide choice. You can use one type of pipe for the entire job, or mix the types to reduce costs or meet specific needs. If the local code limits you to particular types, of course, you'll have to use them. But if regulations in some other area ban a certain type of pipe, don't assume it isn't wise to use it in your own locality. Rules of this type may be based on sound practical considerations, but they are made by human beings who can have strong personal preferences. So they are not always necessarily explainable on a scientific basis.

In choosing your plumbing materials you'll also have to contend with some widely circulated but misleading tales. One that has found its way into print, for example, tells you flatly that you should not use galvanized steel pipe under ground or buried in concrete. Yet even ordinary steel pipe, installed more than half a century ago, is still supplying water to thousands of homes across the continent. Buried in concrete, it carries the hot antifreeze solutions that melt snow from parking areas around some of the world's most modern buildings.

You may also hear unpleasant tales about the effect of copper pipe on drinking water. Yet the world's greatest chefs are accustomed to copper cooking utensils — kitchen items so long established that even Moses referred to them. Also, copper, like iron, is one of the essential elements of the human diet, but it's easier to get the necessary requirements from food than from plumbing. You'd have to drink approximately two buckets of water from a copper household water-supply system to get as much copper as you'd acquire from a dozen oysters.

You'll hear similar scare stories about plastic pipe, fiber pipe, even concrete tile and ceramic tile. But they're in use from coast to coast, and doing nicely. If you're in doubt, contact the manufacturer directly. And keep in mind that our ancestors of only a generation or two ago didn't eat tomatoes because they considered them poisonous.

COPPER AND BRASS PIPE. For all practical purposes, copper and brass pipe are the same size — internal diameter, wall thickness and external diameter — as galvanized and black (nongalvanized) steel pipe. The fittings are also identical, though they should not be mixed, as copper joined to steel results in rapid galvanic corrosion. However, in an emergency, you can use copper or brass fittings with steel pipe or vice versa. Standard pipe-thread dies may be used on copper and brass pipe.

COPPER TUBING. The walls of copper tubing are always thinner than those of pipe; thus tubing has a smaller external diameter than the same size pipe. Except for very small sizes, both tubing and pipe are sized according to their internal diameters. Thus the internal diameter of 1-inch tubing is close to that of 1-inch pipe.

Three types of copper tubing are generally used for home plumbing: rigid tubing, soft or flexible tubing and DWV tubing. Rigid and soft tubing are identical in dimensions. The rigid is tempered and comes in straight lengths. The soft is soft and comes in coils. The rigid looks better when exposed because the soft always retains some of the initial curl no matter how carefully you try to straighten it. Both types are cut and joined exactly the same way. However, you will find it easier to make flared joints in soft tubing, and soft tubing can be hand bent with the aid of a spring bender. Rigid tubing requires a lever bender, which is far more costly than the spring type. As previously stated, the temper is easily removed from rigid tubing with the aid of a propane torch.

DWV tubing is rigid copper tubing made especially for drain, waste and vent piping. It differs from other tubing only in that its walls are thinner, which lowers its cost. It may be cut and joined exactly like any other copper tubing.

Making a soldered joint is a simple matter. To do it right, first clean the end of the tube on the outside and the fitting to be connected on the inside. You can do this quickly with the very fine abrasive cloth used by auto-body finishers (available from auto-supply dealers). Clean the parts just enough to make them shiny, and don't overdo it. The tube and fittings are made to close tolerances. Too much abrasive work increases the gap between them and can result in leaky joints.

Use an old toothbrush to spread soldering flux (a thick paste) over the cleaned areas on the outside of the tube end and the inside of the fitting. Then slip them together. You don't need to apply any force for this.

COPPER WATER-SUPPLY TUBE – TYPE L MEDIUM WEIGHT

Nominal Size	Nominal Dimensions Outside Diameter	Inside Diameter
1/4"	.375"	.315"
3/8"	.500"	.430"
1/2"	.625"	.545"
5/8"	.750"	.666"
3/4"	.875"	.785"
1 "	1.125"	1.025"
1 1/4"	1.375"	1.265"
1 1/2"	1.625"	1.505"
2 "	2.125"	1.985"

The handiest tool for heating the joint to solder-flow temperature (unless you have a gas welding outfit) is a propane torch. This is lightweight and clean. Play the torch flames on the fitting, touching the solder against it at intervals until you see that it's hot enough to melt the solder. Be sure to play the flame also on the tube, close to the fitting. But *do not* point the flame into the joint. (Use solid wire solder that comes on spools, and buy it where you buy your tube.) When the joint is hot enough the solder will suddenly melt at the tip of the solder wire and flow into the space between pipe and fitting by capillary action. The first time you do it you'll probably be surprised. It works just as well even when the solder has to flow upward vertically. Keep pushing the solder wire against the hot joint until you can see a gleam of solder all the way around the seam, which tells you the joint is filled. As the heat stays in the joint for a while, you'll have time to wipe off excess solder with a rag — but don't burn your fingers. Then let the joint cool while you get on with your work. If you pour cold water on a

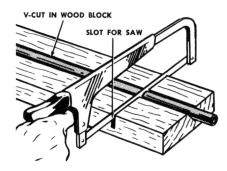

V-CUT IN WOOD BLOCK

SLOT FOR SAW

Cut copper pipe with hacksaw, using a jig to insure a square cut. Jig is simply a wooden board with a V-notch to hold the pipe and a slot for the saw at right angles to the notch.

After reaming inside of cut pipe and cleaning outside with fine abrasive cloth or steel wool, coat inside of fitting and outside of pipe with soldering flux. Then slip fitting on and rotate it several times to spread flux.

To solder connection, heat fitting and pipe with a torch, then apply wire solder from spool to seam between fitting and tube. Do not play the flame on the solder. Remove flame as soon as the seam is filled all the way around.

joint you may crack the fitting, but after it has cooled for a few minutes you can wet it to bring it to handling temperature.

After you've completed a soldered joint, whether small or large, the best test of your work is provided when you turn on the water, assuming you've been working on water-supply pipes. On drain pipes, if it's possible to block the outflow at the bottom of the drain line, you can test by filling the upper length of the pipe with water.

Leaks are not likely, but when they occur they're not difficult to cure. For the remedy, first drain *all* the water from the pipe. Then reheat the joint to soldering temperature and flow in more solder. During this and all previous steps in making solder pipe joints, avoid overheating the paste flux in the joint so it sizzles and smokes. (A little sizzling isn't serious.) Too much heat destroys the effectiveness of the flux, which is intended to assure a thorough and complete bond between the solder and the metal of the pipe and fitting.

For pipes larger than 2″ you may need two torches, as copper conducts heat away from the flame area very rapidly. One final tip: always play the torch flame on the pipe and the fitting, but *not* on the solder.

To make a flared connection you need a flaring tool and a flare type of fitting matched to the tube size. For your first step, slip the flange nut from the fitting onto the end of the tube, and push it back an inch or so. *Then* flare out the end of the tube with the flaring tool. Next, press the flared end of the tube against the polished end of the fitting, and tighten the flange nut onto the fitting. This nut draws the flare snugly against the matching fitting surface, and the job's done. You can buy several types of inexpensive flaring tools at any hardware store.

Making a compression fitting connection is even simpler. The flange nut (shorter than those for flare fittings) is slipped on the end of the tube first, as with a flare fitting. Then the compression ring is slipped on. This looks like a brass wedding ring matched to the pipe diameter. To complete the job simply push the end of the tube into the fitting as far as it will go, and tighten the nut onto the fitting. This squeezes the ring tightly onto the tube and simultaneously seals it against the fitting. That's all there is to it. The tubing should be cut with a tubing cutter for both flared and compression type of connections to assure a perfectly smooth and squared cut.

Flaring tool consists of a vise to hold pipes of different sizes and a ram that screws into pipe end and flares it.

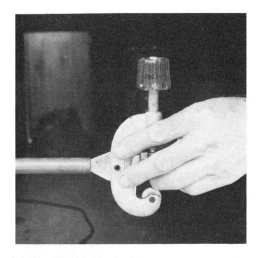

1. Remove the internal burrs at the end of the tube. Here the reamer on the tubing cutter is being used. Then slip the flange nut over the end.

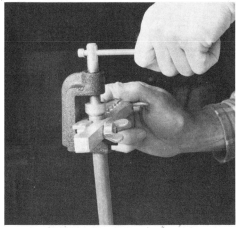

2. Clamp the tube into the vise of the flaring tool, then screw the ram into the end.

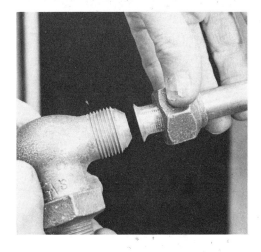

3. Place the flared end against the end of the flare fitting and screw on flange nut.

A flexible-ring compression joint. When the parts of this joint are in good condition, the flange nut need be no more than finger-tight to hold water.

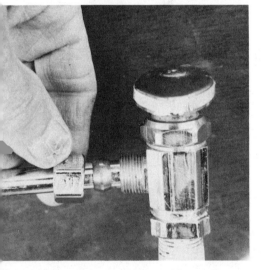

A metal-ring compression joint. Tube must be perfectly round, straight and free of external burrs, otherwise it will leak.

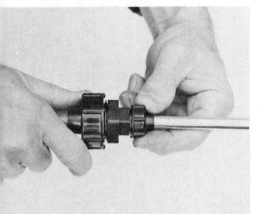

Click & Seal fitting can be used for quickly joining copper and plastic tubing without tools. Coupling shown connects two different tube sizes.

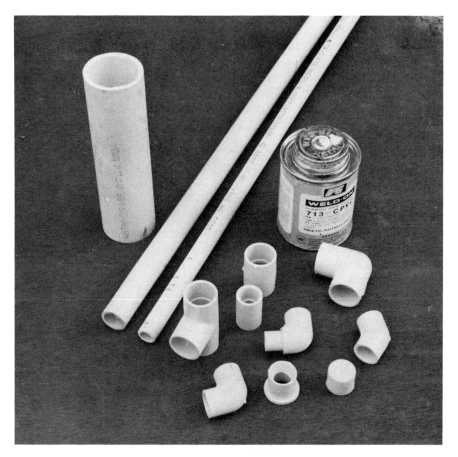

A few examples of plastic pipe, plastic fittings, and solvent cement.

PLASTIC PIPE, though widely used in industry, is not yet common in household plumbing, largely because local codes are slow to approve new materials. Also, at present, only a few types can withstand hot-water system temperatures.

The advantages of plastic pipe and its fittings are low cost, light weight, ease of assembly, and high resistance to corrosive elements that can damage metal pipe. In its flexible form (polyethylene) it has been widely used for quite a few years in deep wells. Its light weight and flexibility plus the fact that it is available in continuous lengths as great as 1,000 feet make it an excellent choice for well use. It can be lowered into the well and removed when necessary almost as easily as garden hose.

In rigid form, plastic pipe has been accepted for certain applications by various state and local codes. It is gaining, for example, in waste-system plumbing. If it's approved in your area, and you plan to install the plumbing yourself in a new house, the lightweight plastic can make the large-pipe drainage work much easier.

JOINING PLASTIC PIPE

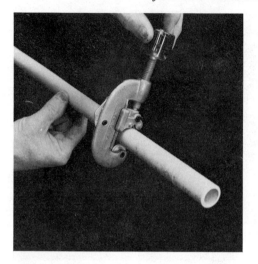

1. Plastic pipe can be cut with a saw, but the job is more easily and accurately done with a standard tubing cutter.

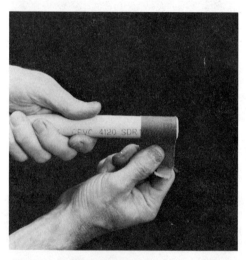

2. To insure good adhesion, remove the shine from the end of plastic pipe with sandpaper or emery cloth.

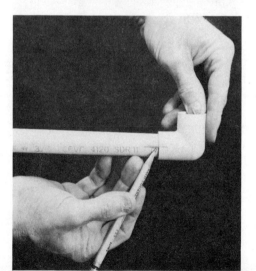

3. Try the fitting to make certain it isn't too loose or too tight. Then, mark the position with a pencil.

 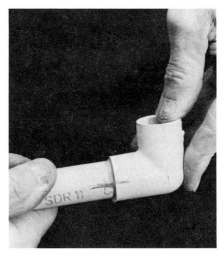

4. Apply a thin, even coat of the solvent to the joint surface and allow to dry. Then spread a thick, even coat of solvent over the joint surface.

5. Press the fitting into place over the pipe end, give it a partial turn and line up the pencil marks. Hold in place until the solvent dries sufficiently for a bond to form.

You can cut plastic pipe with a wood saw, hacksaw, tubing or pipe cutter. A saw is good enough if you use a miter box or other jig to make certain the cut is square and do not have too many cuts to make. The tubing cutter is much faster and you don't have to worry about making a square cut; the tool takes care of that.

Plastic pipe joints. "Solvent welding" is the term used for cementing (gluing) plastic pipe to plastic fittings. The method is fast and dependable. Like a soldered or welded joint, a properly made solvent-welded joint is permanent. Unlike a soldered joint, which can be taken apart by heating and melting the solder, a welded plastic joint can never be taken apart. Once made, the parts form a solid piece of plastic.

To make a plastic joint, start by trying the fitting on the pipe end. It should not fit too tightly. If you have to force the fitting onto the pipe end, there will be no joint or a very poor one because the cement will be driven out. On the other hand, if the fit is so loose you can feel the play, the joint will be weak because there will be too much strain on the cement between the mating surfaces. So try the fit first. If it isn't correct, use another piece of pipe.

Next, use fine sandpaper and remove the shine from the end of the pipe and the interior of the fitting, if you can reach it. Do not touch these areas with your hands if you can possibly help it. Assemble the joint. Rotate the fitting so it points in the proper direction. Use a pencil to mark both the fitting and the pipe so you can quickly reposition the fitting. Then apply the solvent to the joint to be.

You must use the proper solvent; ABC for ABC plastic, PVC for PVC plastic and CPVC solvent for CPVC pipe. And it is best to use a brush that is just as

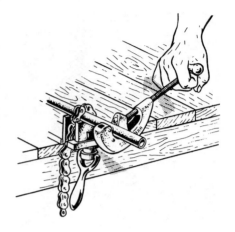

Pipe cutter assures a straight cut with either copper or galvanized pipe. Clamp pipe in a vise with adapter jaws. Hardened cutting wheel is turned deeper into pipe's surface with screw handle after each turn of the cutter.

wide as the band of solvent you plan to apply. The brush must have animal bristles; plastic bristles will dissolve in the solvent. If you don't want to buy a brush, you can make do with the daub provided with the solvent-can cover. The solvent must be at 70 degrees or so and fairly thin. If it is cold, warm it in warm water. If the solvent has thickened, discard it.

Apply the solvent in a *thin* even coat to the end of the pipe and to the interior of the fitting. Wait a minute for the solvent to soften the plastic. Then apply a *thick* but even coat of solvent to the same area. Quickly insert the pipe into the fitting. Give the fitting a quarter turn or so and align the pencil marks. Hold the fitting firmly in place so the cement cannot push the fitting off. Wait a minute. The joint is now made and you can proceed with the following joint. In warm weather you can apply water pressure in a few hours. In cold and to be safe it is best to wait overnight.

If your joint leaks, dry it thoroughly and try forcing solvent into the crack. If that doesn't help, you have to cut the fitting and adjoining pipe out and start all over again.

How to make flexible plastic pipe joints. If you're using *flexible* plastic pipe for a well system, simply slip a metal joint clamp on the end of the plastic pipe, push the end of the pipe as far as it will go *on* (not into) the fitting, and tighten the clamp. Standard fittings include elbows, T's, reducers, and adapters to join plastic pipe to metal pipe. You can cut flexible plastic pipe neatly with a sharp knife or any fine-toothed saw.

GALVANIZED STEEL PIPE costs less than copper and is often available when copper is not. Connections require more work, as the pipe must be threaded after cutting. But steel pipe is much tougher than copper — a factor that can be important where plumbing may be exposed to mechanical damage. In a garage or basement workshop, for instance, plumbing is sometimes subjected to impact, as with tools or heavy objects. If you accidentally strike a copper or plastic pipe with a hatchet you are likely to need a new pipe. If you strike a steel pipe, however, you may need a new hatchet.

To make a steel pipe connection (after threading the cut-to-fit pipe as described in Chapter 2), simply screw the fitting on the pipe, or the pipe into the fitting, as the case may be. The pipe threads, however, should be pre-coated with pipe-joint compound or wrapped with plumbers' teflon tape. The compound (often called Pipe Dope, from a tradename) resembles thick paint. The teflon tape is so thin and elastic that it sinks into the threads almost like a fluid coating when it is drawn tight. Both materials do a double job. They help seal the joint against leakage and they make the joint easier to take apart if it should ever be necessary.

Screw the pipe and fitting together "hand tight" to start, usually about three or four turns. Then complete the tightening with your two stillson wrenches, one on the pipe, the other on the fitting. Place the wrenches so you will do your tightening by pushing the handle *toward* the open side of the jaws. (If the handle is pushed in the opposite direction the jaws won't grip.) Naturally, the two wrenches face in opposite directions. Two or three threads on the pipe will still be visible when the connection is fully tightened. The threads on the pipe end and on the inside of the fitting are tapered to match — so they jam firmly together as the parts are tightened. This, together with the pipe-joint compound or the teflon tape, makes a tight seal — so you don't need brute force to do a good job.

Buying galvanized steel pipe is a simple matter. For big jobs it comes in 21' lengths with both ends threaded and a coupling screwed on *one* end. For very small jobs, you can buy pipe "nipples," commonly up to about 6" in length. And some hardware stores and plumbing suppliers will cut and thread specified lengths to order. (If this service isn't available in your area you can rent the tools to do it yourself.)

STANDARD CONTINUOUS STEEL PIPE

Inside Diameter	Size: Nominal (O.D.)	Outside Diameter
.26″	1/8″	.405″
.36″	1/4″	.540″
.49″	3/8″	.675″
.62″	1/2″	.840″
.82″	3/4″	1.050″
1.04″	1 ″	1.315″
1.38″	1 1/4″	1.660″
1.61″	1 1/2″	1.900″
2.06″	2 ″	2.375″
2.46″	2 1/2″	2.875″
3.06″	3 ″	3.500″
3.54″	3 1/2″	4.000″
4.02″	4 ″	4.500″

COPPER DRAINAGE TUBE (DWV)

Nominal Size	Actual Outside Diameter
1 1/4″	1.375″
1 1/2″	1.625″
2 ″	2.125″
3 ″	3.125″
4 ″	4.125″
5 ″	5.125″
6 ″	6.125″

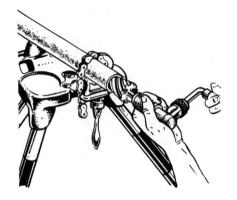

Pipe and tubing ends should be smoothed with pipe reamer in bit brace. A few turns do the job. Tool is also handy for odd jobs like enlarging bored holes.

DRAINAGE PIPE differs from water-supply pipe in a number of ways even though it may be made of the same material.

Drainage fittings have certain special features. Many, for example, are designed to change the direction of the pipeline with a gentle curve rather than a sharp angle so that solid matter passing through the drain doesn't pile up because of a slowing of the flow. Most drain-pipe fittings also are designed so no ridges or pipe shoulders protrude into the flow. The smooth passage through the fitting minimizes the chance of solids snagging. Right-angle elbows are also available machined or threaded to provide an angle just a shade larger than 90 degrees. When you use one of these to connect a vertical pipe to a lateral one (as across a basement ceiling) the lateral pipe is automatically given a downward pitch of ¼" per foot of run. The pitch is barely perceptible but it effectively prevents accumulation of sludge along the lateral run.

Cast-iron drainage pipe has probably been most widely used for soil and waste stacks (See Chapter 1) over the years. It's made in "service weight" and in "extra-heavy" weight, and you can buy it pretreated with a preservative coating or without it. Whatever grade you use it's likely to last at least until your great grandchildren are married.

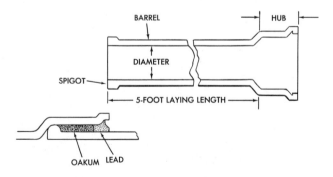

Cast-iron pipe used in main stacks and sewer lines comes with hub at one end, spigot at other. Spigot is inserted in hub, then packed with oakum and leaded, as shown in detail. Caulking iron then spreads lead to tighten it.

Cutting cast-iron pipe. A pipe cutter won't work on it, so you have to use other tools. On standard-weight cast-iron pipe you can make a $1/16''$-deep cut all the way around it with a hacksaw. Then rest one section of the pipe on a piece of 2-by-4 with the saw cut just overhanging. Next, tap the pipe overhang with a hammer close to the cut until the cut section breaks off.

Another method calls for cold-chiseling all the way around the cut, gradually hitting the chisel harder until the pipe separates at the cut. Fortunately, by planning carefully, it's usually possible to minimize the number of cuts required. And, if you're afraid of shattering the pipe, it's not too difficult to saw the standard weight all the way through. A blade with about fourteen teeth per inch does the job fastest.

Lead caulked joints are commonly used to join cast-iron pipe to itself and to any of the numerous cast-iron fittings available. The pipe, usually sold in 5' lengths, has a larger diameter "hub" at one end, and a slight ridge around the other end, called a "spigot." It's also available with the hub at both ends. The hub and spigot diameters are such that the spigot fits inside the hub of an adjoining section with space to spare. And that extra space is the basis for the joining system.

Whether you're joining one length of cast-iron pipe to another or to a fitting, the method is the same. In vertical runs of pipe, like the soil stack, the pipe is always erected with the hub ends *up*, so the lower end of each section of pipe fits inside the hub of the one below it.

When two sections are fitted together make sure the upper one is *centered* in the hub of the lower one. Then pack "oakum" down into the space between them. This is a stringy fiber available from your plumbing supplier.

The usual tool for the packing job is a "yarning iron" made for the purpose, or a blunt-ended piece of narrow strap iron about a foot long. Simply wrap the oakum loosely around the inner pipe and bang it down firmly with the tool, using moderate hammer taps to pack it. Usually, you pack the oakum to within $3/4''$ or $1''$ of the top of the hub. (Your local plumbing code, if there is one, will specify this.)

The next step is to pour molten lead into the hub on top of the oakum. Melt the lead in an iron pot available at hardware stores or plumbing suppliers. Professional plumbers heat the pot on a plumber's furnace — which is really a propane blowtorch that shoots its flame upward and serves as a support for the pot. You also need an iron ladle (made for the purpose) to scoop the lead out of the pot and pour it into the joints. And you need heavy work gloves to protect your hands when the ladle handle heats up. The pot and the ladle are inexpensive, however, and the whole job is really easier than it sounds. (It's important to heat the ladle in the torch flame briefly before filling it, as even a slight trace of moisture in it can cause the lead to spatter dangerously.)

As soon as the lead has hardened it is caulked with a pair of tools called caulking irons. These resemble offset chisels. The "inside" iron is shaped to make it easy to pack the lead against the pipe. The "outside iron" is designed to pack it against the inside of the hub. You do the packing by simply driving the sharp edge of the tool into the lead to spread it. You'll usually have to work the tools around the joint several times to tighten the joint completely.

To cut cast-iron pipe start by making $1/16$"-deep cut with hacksaw all the way around, and be sure the cut is squared with pipe. Chalk or crayon line around pipe is good guide.

If pipe is standard weight you can break off end beyond hacksaw cut simply by following up with hammer taps all around until end clicks off.

If pipe is extra heavy grade follow hacksaw cut with cold chisel, tapping it into cut all the way around. Keep turning pipe and tapping chisel into cut until end breaks off.

Push oakum into hub after pipe sections are fitted together, and pack it in tightly with yarning iron. Buy the iron in the same place you buy the pipe.

Molten lead poured into hub on top of packed oakum is next step in sealing joint.

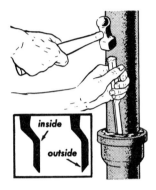

Tap caulking iron into cooled lead to spread it against pipe surfaces. Use "inside" iron to spread lead against surface (inner) of hub. Use "outside" iron to spread it against outer surface of pipe fitted into hub.

To make a horizontal poured-lead joint, begin as usual, with oakum packing. Then fasten "joint runner" around pipe snug against open end of bell, like this. Pour molten lead into opening at top of joint runner. Then cut off lead bump left at pour-in opening, using hacksaw with coarse teeth. Finish with caulking irons, as for vertical joints.

Making a horizontal joint in cast-iron pipe calls for a special gadget called a joint runner. This is simply a fat asbestos rope with a clamp to hold the ends together. After packing the joint with oakum, wrap the runner around the pipe, clamp the ends together at the top, and push it up against the joint tightly. If it isn't snug, tap it in lightly with a hammer. The clamp is designed to leave a little space at the top and serve as a funnel when you pour the molten lead. Keep your feet out of the way in case you spill any. When the joint is filled, you'll see the lead rise in the upper space. After it cools you can chisel off the bump of lead that remains on top where the clamp was. Then caulk the joint with the caulking irons just as in the vertical joint.

No-hub joints are another way of joining cast-iron pipe to itself and to cast-iron fittings. The No-hub joint is a newly developed fitting consisting of a neo-

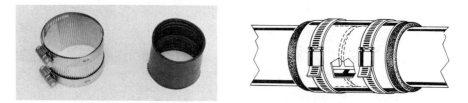

A No-hub fitting (left), showing the flexible neoprene sleeve and stainless-steel metal sheet and holding clamps. The sleeve slips over the joint and is held by the sheet and clamps (right).

prene sleeve into which the pipe ends, without ribs or hubs, are positioned, a flexible stainless-steel metal sheet that is wrapped around the neoprene sleeve, and two stainless-steel clamps that hold everything together. The No-hub is several times easier and faster to make than any other cast-iron joint. It can also be easily taken apart without damage, something that is difficult to do with caulked lead joints. And the Non-hub joint can be bent a few degrees without leaking—again, something that is impossible with caulked lead joints.

No-hubs can be used with standard cast-iron pipe and fittings after the ridges and hubs have been removed; they can be used with cast-iron pipe and fittings manufactured especially for use with No-hubs and with all, non-threaded pipe, including plastic.

FIBER PIPE, VITRIFIED TILE, AND CEMENT TILE. These are used in the outdoor portion of your drainage system, along with cast iron and copper. (Your local code will either specify the type or give you a choice.) Usually, the rules vary according to the type of disposal system. You may have a wider choice if you are using a septic tank than if you are connecting a municipal sewer.

Fiber pipe (like Orangeburg pipe) is in the higher end of the price range, but it's a great work saver. It's sold in various lengths, some of the common ones being 8' and 10'. The better grades are made with coal-tar formulas which are unaffected by detergents. But be sure the pipe you buy is suited to the use you plan for it. You can check this with your plumbing-supply dealer—and, of course, your local code.

Two types of fiber pipe are in common use. One is the conventional form used to convey drainage material *to* the disposal system. This is similar to other types of pipe. The other form is designed to permit the liquid effluent from septic tanks to escape into the surrounding soil to be absorbed. This type is perforated with small, evenly spaced holes in straight rows, usually on opposite sides of the pipe. The same fittings are used in connecting the perforated sections as the standard ones.

As both forms of fiber pipe are made in sections of much greater length than vitreous and cement forms, they make it easier to maintain a constant downward pitch of the required degree when the pipe is laid in trenches, as for connection to a sewer or septic tank. The usual pitch of ¼" per foot, for example, is so slight that minor variations in grading can upset it unless considerable care is used with shorter sections of pipe. The pitch of the pipes in drainage fields (where

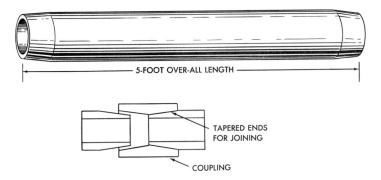

Fiber pipe lengths are connected by tapered couplings.

fluids are released for earth-absorption) is often as little as ⅛" per foot, so variations can be even more important. (More about this in Chapter 8, Septic Tanks and Drainage Fields.)

Connecting fiber pipe. In its usual form, both ends of each length of fiber pipe are tapered to match the internal taper of couplings and fittings. To connect a fitting you simply push it on the end of the pipe by hand as far as it will go, then hold a wood block across the fitting end, and hammer-tap the block to drive it on all the way. The friction of the drive-on actually fuses elements of the coal tar, sealing the joint.

To cut fiber pipe, use a coarse-tooth saw like the larger frameless hacksaws. (Fine teeth clog with the pipe material.) Be sure to cut the pipe squarely. But remember that cutting removes the tapered end.

Joining cut fiber pipe. Even though they have no taper, ends can be joined by using special fittings made for the purpose. The nature of these fittings varies with the manufacturer. In some brands, a special coupling is used to splice the tapered end back onto the pipe so the usual method can be followed to connect the overall length. Other makes employ a neoprene outer sleeve and metal band clamps to join cut ends.

Fittings include elbows, couplings, special couplings (for cut sections), Y branches, and crosses. Special types of crosses are made for septic-tank drainage systems by some manufacturers.

Vitrified tile pipe is dark, reddish-brown, with a shiny surface. You often see it stacked in building-supply yards. It is made with a hub at one end, like cast iron pipe, but the other end does not have the shallow ridge found on cast-iron pipe. At most supply outlets you'll find an assortment of fittings almost as large as for cast iron, though types for connecting to smaller diameters than the usual exterior drainage sizes are not common. And you don't need them.

To make a vitrified pipe joint, fit the small end of one section into the enlarged hub end of the next one. Then, holding or blocking up the joint to keep it centered, tamp oakum in the space between the inside of the hub and the outside of the connecting pipe. Don't use a yarning iron for this, however, because of the pipe's brittleness. A thin piece of wood, like a planed-down lath or lattice strip makes a good substitute. Tamp in just enough oakum to seal the joint so that the masonry cement that completes it won't ooze into the pipe. (Hardened little snags of cement inside the pipe can cause clogging later.)

Unless your local code specifies otherwise, use a mix of one part Portland cement and two parts clean mason's sand. Add just enough water to get an easily workable putty-like consistency. Although individual sections of vitrified (often called glazed) tile were once limited to 2', you can now get them from many outlets in 3' and 4' lengths. If you have only 6' or so to connect you can make the job simple by propping the tiles in vertical position for the cement job. Then, after the cement hardens, you can lay them in the ground. On longer runs, lay several layers of newspaper under the joint to keep loose earth from mixing with the cement while you work. You'll have to do some exercise to get the oakum tamped in, but it's not too difficult. You can trowel the cement in on the top and sides of the joints, but you'll probably find it easier to use your fingers to push it in on the under side. Despite the glass-like surface of the pipe, the hardened cement gets such a solid grip you can't separate the joint without breaking the pipe.

You're likely to be using this type of pipe to connect your house drainage system to a septic tank or sewer. Codes vary on the types permitted, usually specifying cast iron for a certain distance outward from the foundation wall, then permitting glazed tile. The common diameter for these outside drain or sewer lines is 4" — measured inside the pipe.

Concrete tile pipe (sometimes called cement tile) is usually the lowest-priced pipe for septic-tank drainage fields. Whether you can use it or not depends on your local code. It has established a very good record for durability over quite a few years. As drainage field tiles are intended to allow the watery effluent from the septic tank to seep into the earth over a wide absorption area, they are not connected like the tiles carrying the sewage to the tank. Instead, they are simply laid end-to-end with 1/8" to 1/2" space between them to allow the effluent to leach out into the ground. If your local code doesn't specify the space between tiles, 1/4" usually works out well. Tar paper or roofing paper (in small rectangles) is laid over the top of each joint to prevent earth from sifting through the openings when the tiles are covered.

Clay tile, kilk hardened like pottery, is another type used in drainage fields. This is yellowish in color in some forms, red in others. It costs more than concrete tile, but does its job well. It is laid in the same manner and with the same spacing. Both types are available in 1' lengths of 4" diameter.

UNTANGLING PIPE SIZES. The diameters of most pipes are expressed in "nominal" sizes like the dimensions of stock lumber. If you've bought much lumber you probably know that a 2-by-4 is really only about 1 5/8" by 3 5/8", and that other sizes vary from the nominal in the same way. The actual lumber size is always a little smaller than the nominal. The nominal diameter of standard-weight pipe and tubing, however (always based on the *inside* diameter), sometimes works the other way. The actual size of the pipe may be greater than the nominal size. But which way it works depends on the size of the pipe. The inside diameter of typical ½″ copper water tube, for example, is *more* than ½″. But the inside diameter of 2″ copper water tube is *less* than 2″. (The differences are even more confusing with steel pipe. The inside diameter of what is called ⅛″ steel pipe is actually considerably more than ¼″.) The nominal and actual sizes of the commonly used diameters are given in the charts for both copper and steel. Plastic is usually figured on the steel pipe size basis. But if the size factor is critical, it's best to doublecheck the brand, in case of variation.

There's one consoling point about the perplexing pipe and tube sizes, however: you seldom need to know the *actual* inside diameter. The *nominal* sizes recommended later for different parts of the plumbing system tell the story. And the water flow rates mentioned in the other chapters are likewise based on nominal sizes. The *actual outside diameters* listed in the comparison charts, however, may sometimes be useful where holes must be bored or notches cut to take pipe or tubing.

PIPE FITTINGS are made in a wide variety of forms, as shown in the illustrations. They're designed to let your plumbing take right-angle corners, 45-degree turns, and in some cases, turns as slight as 5⅝ degrees. There are also T fittings for connecting a branch pipe into a straight run, and Y fittings for branches that connect at an angle. Where four pipes connect at right angles to each other the fitting is called a "cross." A wide assortment of reducing fittings are also available for connecting small pipes to larger ones along straight runs or at turns or branches.

Generally, you'll be able to buy a fitting for just about every plumbing situation you're likely to encounter with a given type of pipe. But you won't find illogical fittings. You won't find a wide assortment of angle fittings, for example, for flexible plastic well pipe because the pipe can easily bend enough to make the difference.

In addition to the conventional fittings shown in the illustrations, there are numerous special types for unusual situations. If you have a plumbing problem you can't solve with the regular fittings, tell your plumbing supplier what you want to do. He may stock an off-beat fitting that will do the trick — or, more likely, suggest a way you haven't thought of for doing the job with conventional ones. Specific details on using fittings are given in subsequent chapters where they apply.

What a union does. This is a very important fitting in all work done with threaded pipe. The easiest way to understand its purpose is by visualizing some common plumbing situations. Suppose, for example, you have an old, leaky section of pipe running horizontally near the floor of a first-floor storeroom. An elbow at one end of it connects a pipe leading down through the floor to the basement. An elbow at the other end connects it to a pipe leading up through the ceiling to the floor above. How can you remove the horizontal section of pipe?

If you think about it you'll realize that when you try to unscrew the horizontal pipe from the elbow at one end you simultaneously screw it deeper into the elbow at the other end, and it's already as far in as it will go. So the plain fact is that you can't unscrew the horizontal pipe. To get it out by the unscrewing

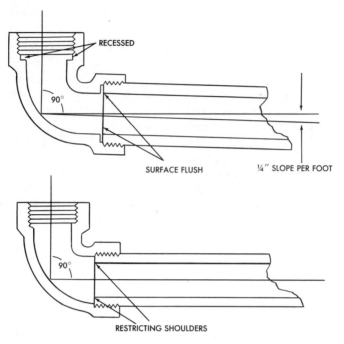

RECESSED

90°

SURFACE FLUSH   ¼″ SLOPE PER FOOT

90°

RESTRICTING SHOULDERS

Difference between drainage fittings (top) and water-supply fittings (bottom). Drainage fittings have no shoulder or bump protruding into fitting where pipe end screws in so as not to catch solid matter. Also, right-angle elbow isn't true right angle, but allows for ¼″ per foot downward pitch of pipe. Water-pipe elbow is true 90-degree angle. Pipe end protrudes slightly inside fitting, forming elbow or bump. As solid matter isn't traveling through water pipe, shoulder doesn't matter.

method you'd have to disassemble all the pipe either above it or below it—turning a simple repair job into a major undertaking. So you merely saw through the horizontal pipe. Then you can turn each of the severed ends in the proper direction to unscrew them from their respective elbows. That lets you remove your leaky pipe.

But how do you put in the new pipe? It won't do you much good if you have to saw it in half. And you can't screw it into one elbow without unscrewing it from the other. That's where the union comes in. Actually you *do* cut the new pipe in half. In fact, you cut off a little extra to make room for the union. Then you thread the severed ends and screw half of the union on to each one, sealing and tightening, as described earlier. (The two halves of the union are separated by simply unscrewing the outer rim that holds them together.) The outer rim remains loose on one half of the union. Next you screw each half of the new pipe into its proper elbow, and tighten it all up. When you tighten the outer rim of the union (it's like a big hex nut), you seal the two halves of the union together without turning either section of the new pipe. That's just one of the otherwise impossible jobs you can do with the little fitting called a union.

You need a union for many threaded-pipe situations. (You can make soldered-

SANITARY TEE

CLOSET FLANGE

CLOSET BEND

DRUM TRAP WITH COVER

45° Y-BRANCH

COUPLING

¼-BEND

STEEL PIPE ADAPTER

⅛-BEND

CLEANOUT WITH PLUG

SANITARY TEE WITH SIDE INLET

SLIP COUPLING

CAST-IRON PIPE ADAPTER

90° STREET ELBOW

Copper Pipe Fittings

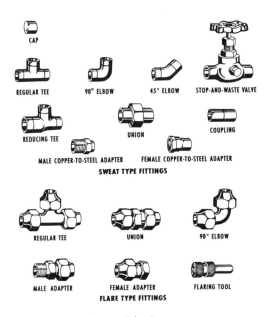

CAP

REGULAR TEE

90° ELBOW

45° ELBOW

STOP-AND-WASTE VALVE

REDUCING TEE

UNION

COUPLING

MALE COPPER-TO-STEEL ADAPTER

FEMALE COPPER-TO-STEEL ADAPTER

SWEAT TYPE FITTINGS

REGULAR TEE

UNION

90° ELBOW

MALE ADAPTER

FEMALE ADAPTER

FLARING TOOL

FLARE TYPE FITTINGS

Copper Tube Fittings

REGULAR TEE

90° ELBOW

45° ELBOW

90° STREET ELBOW

REDUCING TEE

UNION

REDUCER

COUPLING

HOSE ADAPTER

BUSHING

PLUG

CAP

Galvanized Steel Pipe Fittings

Transition fittings or adapters are designed to connect two different kinds of pipe. This fitting connects copper tubing by means of a soldered joint to male-threaded pipe.

Fitting used to connect copper tubing to female-threaded pipe or fitting.

Fitting used to connect plastic pipe to female-threaded pipe or fitting. Fitting comes apart at O-ring.

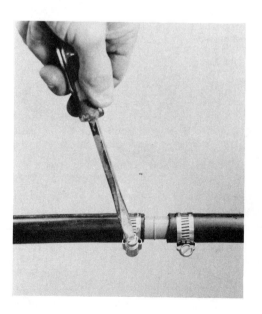

Fitting used to connect two lengths of plastic pipe. Loosening clamps permits joint to be disassembled.

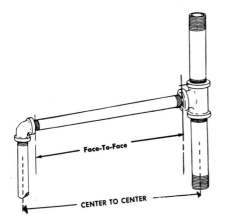

Measure threaded pipe like this. Measure center-to-center for boring holes through which pipe will run, as to faucet connection on floor above. Measure face-to-face distance the threads will screw into the fitting at each end. This means that you add double the distance when pipe goes into fitting at both ends — as it usually does.

joint changes without it, as described in Chapter 5.) If you're installing new plumbing with threaded pipe it's wise to include a union wherever you may have to take a section of piping apart in the future. And, while you're at it, if you have any ideas of future expansion, you can make it easier. At logical points in your water pipes you can use a T instead of an elbow, and close the extra branch of the T with a plug. Then, when you want to extend your water line, all you need do is remove the plug and connect the new pipe. And, more important, if you have an extra bathroom in mind for the future, provide the soil pipe connection in your original installation — if your local code permits it. The connection, of course, must be capped. Connecting into a soil pipe later can otherwise be a tough job.

Transition fittings or adapters. These are fittings designed to connect two different kinds of pipe. For example, to go from copper tubing to galvanized steel pipe a transition fitting is used. With copper tubing you have a choice. The end that connects to galvanized pipe is threaded, but the other end, the tube end, may terminate in a sweated (soldered) joint or a compression joint. To join galvalized to plastic pipe you have a choice of two fittings. Both have threads on the galvanized ends. The other end may be designed to be solvent welded to the plastic pipe or joined by a flexible-ring compression joint.

HOW TO MEASURE PIPING. In order to calculate the amount of piping you need for a particular job, you have to know how to measure. First, take the "face-to-face" measurement from the face of the fitting (the entrance to the opening into which the pipe screws or solders) at one end to the face of the fitting at the other end. Then add the distance the pipe extends into the fittings. Be sure to include the amount for both ends. Next, allow for the space the fittings add at the ends of the pipe. (Some elbows, for example, are longer than others.) The chart gives you the typical screw-in distances and fitting dimensions for common water and drain pipe sizes.

If you don't like arithmetic, you can often do your plumbing on a "dry run" basis — especially installation work. Just measure from the connection where your run of pipe will begin to the point where a fitting is to be located. Then have somebody hold the fitting right on the spot — even if you have to use a cigar box

| Pipe Size | Distance Pipe Is Screwed Into Fitting | |
	Standard Fitting	Drainage Fitting
1/2"	1/2"	—
3/4"	1/2"	—
1 "	5/8"	—
1 1/4"	5/8"	5/8"
1 1/2"	5/8"	5/8"
2 "	3/4"	5/8"
3 "	—	7/8"
4 "	—	1 "

to space it out from the wall, or down from the basement ceiling. This way, all you have to do is add on the screw-in distance at each end. You can simply pencil-mark the desired location of the fittings in your own way, on wall, floor, or ceiling. This may seem like a childish approach—but childish methods that come out right are better than arithmetic that doesn't. Only you can decide which method suits you best. And even the pros make mistakes.

4 | Your Source of Water

THE LOCATION of your home usually determines whether you receive your water from a municipal supply system or from your own well. Which is better depends on the local conditions. If extensive chemical treatment is employed to remove harmful bacteria from a municipal system you may not enjoy the taste. But it can't be avoided. In other areas the addition of various fluorides to the water in municipal systems is compulsory. If you have any physical condition that causes you to react unfavorably to these chemicals you must, of course, use water from another source for your drinking and cooling. Or you must install equipment to remove the fluorides.

The quality of well water, too, depends on the area—and the well. In open country a "shallow" or "dug" well may provide pure and pleasant-tasting water. This is simply a hole, usually from 15' to 30' deep, lined with uncemented stones. Water from rain and other surface sources filters down through the soil and seeps into the well.

In more densely populated areas or in localities where there is very little surface water, a "deep" well is the answer. This type is machine-drilled to a depth of from 100' or so to several hundred, depending on the level at which an adequate water-bearing strata is encountered. This is usually below an impervious layer of rock or other firm material that serves to seal off the deep water from the surface water and its contaminants. And although the deep water supply itself is actually fed from the surface, it undergoes very considerable natural filtration in the process.

To prevent unfiltered surface water from leaking into a deep well a well casing of wrought-iron pipe, usually 6" or 8" in diameter, is driven solidly into the firm layer above the deep water. One or more smaller pipes are led down inside the casing to carry the water pumped from the well. In the relatively rare "artesian" well no pump is needed because the deep water is under enough pressure to drive it up and out of the well. This phenomenon results when a well is drilled into a deep-water strata that receives its supply from much higher ground, such as surrounding hills.

HOW WELL PUMPS WORK. The easiest way to understand the operation of the different types of well pumps is by recalling a few points of elementary physics.

The common soda straw, mentioned in Chapter 1, can serve again to illustrate a basic principle. When you suck water up through a straw you aren't really "pulling" it; the water is actually being "pushed." The atmosphere around you is bearing down on the surface of the water (and everything else) with a constant pressure of about 15 pounds per square inch at sea level. When you create a

47

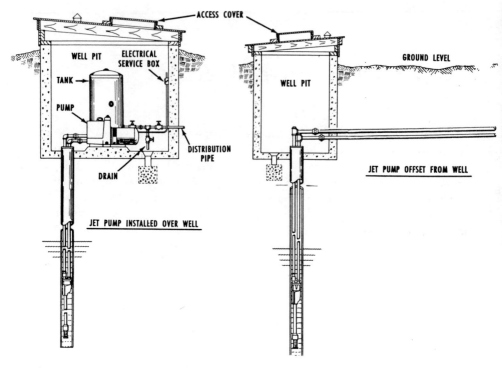

JET PUMP INSTALLED OVER WELL

JET PUMP OFFSET FROM WELL

Jet pumps are the most popular type for deep wells. Two pipes run from pump above ground to jet unit below water level. One drives steam into jet; the other brings back additional water, plus that which was driven down. Pump can be over the well or in the house basement.

lower pressure than this inside the straw, by sucking on it, the pressure of the atmosphere on the water in the glass forces it up the straw. But, as the atmosphere's pressure of 15 pounds per square inch isn't very powerful, there's a limit to how far upward it can push a column of water. The limit is around 33'. At that point the weight of the water in the column exerts enough downward pressure (on a per square inch basis) to cancel out the atmospheric pressure. All the suction pumps in the world can't pull it any higher because the atmosphere simply can't push it any higher. And in practical use, suction types of well pumps seldom lift water from a depth of more than 25'.

The old-fashioned lift pump is the simplest example of a suction type of well pump. (You can still buy this type for camps and summer cottages.) It is merely a smooth-bored iron cylinder with a leather-rimmed piston in it and a handle to move the piston up and down. A spout projects from the cylinder near the top. A pipe from the bottom of the cylinder leads down into the well water. Today, the lower end of the pipe usually has a "foot valve" screwed on it. This works something like the tire valves on your car. It lets water come into the bottom of the pipe but it won't let it run out. So, once you have filled the pipe and the pump with water you don't have to prime it again.

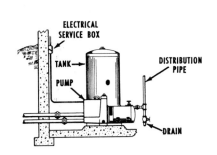

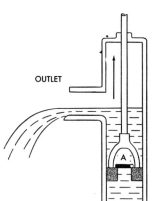

How a lift pump works: When piston is pulled up, weight of the water above it holds valve A closed. Water above piston is lifted to the outlet and flows out. At the same time, the piston, by its upward stroke, has lowered pressure inside the cylinder below. Atmospheric pressure, pushing downward on the water in the well, shoves the water into the inlet (toward low pressure area) through foot valve C and so on up through lower pump valve B. When piston comes down again, it pushes valves B and C closed, opens valve A to let water through. Up stroke then repeats the cycle.

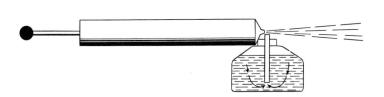

Bug sprayer and jet pump operate on the same principle, Bernoulli's Law. In the sprayer, the high-velocity air stream lowers air pressure directly above the tube which extends into the insecticide reservoir, causing the insecticide fluid to rise in the tube from where it is atomized by the air stream.

When the pump piston is pulled up from the bottom of the cylinder it "sucks" water up the pipe, forcing open a one-way valve (like the foot valve) in the bottom of the cylinder. And more water pushes into the pipe through the foot valve. The water that was already in the pump cylinder above the piston is lifted up and out of the spout to your old oaken bucket. The weight of the water remaining in the pump and pipe closes the valve in the base of the pump and the foot valve below, so the system stays full and nothing runs back into the well. When the piston is pushed down again, a one-way valve in the piston itself opens to let water flow upward through the piston as it descends. Once the piston reaches bottom and starts up again, this valve closes so the water above the piston is lifted on the upstroke.

Today's automatic shallow-well piston pumps are essentially motorized versions of the old lift pump. They're sealed, however, so they can build up some pressure in a water-storage tank. To maintain the pressure, a small amount of air is bubbled in along with the water. As the water rises in the tank, it compresses the air above it, maintaining a steady pressure in the water-supply system even when the pump is automatically shut off by the "pressure switch." This switch, about the size of an orange, is connected to the pump by a slim copper tube. As the water pressure in the tube increases it pushes a flexible diaphragm in the switch outward until the switch finally clicks off. Then, as the pressure decreases from the use of water in the house, the diaphragm flexes in the opposite direction until the switch clicks on again. The same type of switch is used on most other types of automatic well pumps.

The deep-well piston pump was the first type devised to bring water up from depths too great for suction pumps. It's seen mainly in old installations today, but it illustrates an important point: If the pump itself is under the water in the well, it can *push* the water upward almost any reasonable distance, even a hundred feet or more. So the deep-well piston unit was at the bottom of the pipe rather than at the top. A long, wooden piston rod, joined in sections to match the coupled lengths of pipe, dangled down the center of the pipe from a motor-driven crank unit above the well. The buoyancy of the wood made it practically weightless in the water-filled pipe, so a relatively low-powered motor could raise and lower the piston even a hundred feet or more below. The pump and valves were essentially the same as those of the suction system. But the main drawback was the difficulty of repairing the pump. The whole rig had to be hoisted out of the well with block and tackle, and dismantled a section at a time just to get at the pump.

The jet pump, now one of the most popular types, brought a new principle to deep-well pumping—Bernoulli's Law. In greatly simplified terms, this tells you that the outward pressure of a fast-moving stream of gas or fluid decreases as its velocity increases, and becomes much lower than the pressure of the stationary gas or fluid around it. Thus, the higher-pressured stationary gas or fluid tends to be sucked toward the fast moving stream, and into it. Whenever you use a pump-type bug sprayer you see a demonstration of this. If you look at the front end of the sprayer you'll see a little hole where the air from the hand pump shoots out over the open top of the tube leading into the insecticide container.

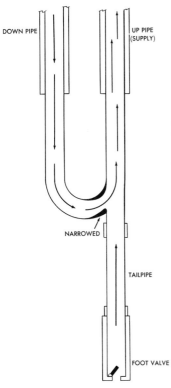

DOWN PIPE

UP PIPE
(SUPPLY)

NARROWED

TAILPIPE

FOOT VALVE

Jet unit beneath well water utilizes fact that high-velocity water stream lowers pressure, tends to draw adjacent water into it. Downshot water stream is speeded up by narrowed section (Venturi tube). To get through this it must travel faster —like water coming through the narrow nozzle of a hose. High speed lowers outward pressure below atmospheric pressure, so well water is sucked into stream and more water goes up than came down.

The low pressure of the high-speed air stream draws the insecticide up the tube—with atmospheric pressure pushing it (see diagram).

The jet well pump uses the same idea in a surprisingly simple way. It shoots a very fast stream of water *down* a pipe into the well, then around a hairpin bend far below the water's surface, and up again through a return pipe. The hairpin bend, however, has some special and interesting features. One section of it is narrowed to form a "Venturi tube." In order to get through the narrowed section the water must travel faster than in the preceding run of pipe—so its sideward pressure is reduced even more. And this section has an opening connected to a "tail pipe" that admits well water through a foot valve. As the high-speed stream through the narrowed section is well below atmospheric pressure, well water rushes into the tail pipe—pushed by atmospheric pressure—and is carried *up* the return pipe to the pump. So the pump gets more water back than the amount it sent down, and the surplus goes into the water-storage tank. Air is bubbled in and pressure controlled just about the same as in the automatic piston type of pump for shallow wells.

The big advantage of the jet system is the fact that the pump is easily accessible above ground. The hairpin bend with its Venturi has no moving parts likely to break down, so it isn't likely to need hauling up very often. And, as it can be suspended a hundred feet or more on lightweight, flexible plastic pipe,

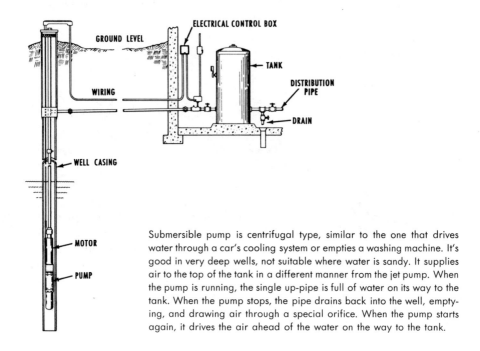

ELECTRICAL CONTROL BOX

GROUND LEVEL

TANK

WIRING

DISTRIBUTION PIPE

WELL CASING

DRAIN

MOTOR

PUMP

Submersible pump is centrifugal type, similar to the one that drives water through a car's cooling system or empties a washing machine. It's good in very deep wells, not suitable where water is sandy. It supplies air to the top of the tank in a different manner from the jet pump. When the pump is running, the single up-pipe is full of water on its way to the tank. When the pump stops, the pipe drains back into the well, emptying, and drawing air through a special orifice. When the pump starts again, it drives the air ahead of the water on the way to the tank.

it can be hoisted by hand almost as easily as a garden hose can be reeled in. If the plastic pipe should develop a leak, it's one of the least expensive types, anyway, and can be replaced with a minimum of work.

The submersible pump, now gaining in popularity, can bring water up from even greater depths than the jet, and it requires only one pipe down the well instead of two. But it has a higher initial price. It is simply a long, slim centrifugal pump connected to the lower end of the supply pipe running down into the well. Because the pump tended to twist the pipe in starting, it was originally suspended on metal pipe. Now, however, a semiflexible, lightweight plastic pipe is widely used. This withstands the twist, has the electric cable to the pump embedded in it, and can be hauled up almost as easily as the flexible jet pipes. An advantage of the submersible pump: as it is located down in the well it requires no space above ground, and you can't hear it when it's running.

WATER TREATMENT EQUIPMENT is available to handle practically any water quality problem from simple hard water to bacterial pollution. For example, water softeners are available to treat your water before it enters the supply pipe, removing the minerals that cause washing problems and deposits on cooking utensils. Chlorinators are also a stock item. Many of these use ordinary household bleach, automatically metered into the water supply to destroy bacteria and eliminate the water odors present in some areas. And secondary equipment can be added to remove any remaining traces of chlorine if you require water entirely free of chemical taste or smell. Typical units for water treatment are priced in about the same range as TV sets. And, if you need absolutely pure distilled water, for critical baby formulas or tricky photo processing, you can get it with an automatic distilling unit that produces about 1½ gallons a day while using about the same amount of electric current as a bright work light.

Municipal or state water testing is available in most areas either gratis or at a nominal charge. Manufacturers of water-treatment equipment, too, often offer water tests gratis to determine the type of equipment required. The best way to locate water-treatment specialists is to look in the phone book's Yellow Pages.

SELECTING AND INSTALLING PUMPS. The depth of the water level in your well and the number of gallons per minute the well can supply are factors that affect your pump selection.

Don't try to use a shallow-well pump if your water level is much below 20! Ask the well driller (if it's a new well) how many gallons per minute the well can supply. Well drillers often use a portable pump and tubs or buckets of known capacity to measure a well's output by timing the filling time and measuring any decrease in the well's water level. The object is to avoid a pump installation that might suck the well dry, thus losing the "prime" of the pump system. If this should occur unnoticed, the dry-running pump could be seriously damaged.

Where the water level in a deep well is normally high, the depth of water in the well casing provides a good margin in gallons. As the pump runs only periodically, it can temporarily lower the well's water level while filling the storage tank and bringing the pressure up. But as long as it doesn't drain the well down to the intake footvalve, it will retain its prime (not run the pump dry). When the pump shuts off, this water level in the well will rise again to normal. Your best bet: use a jet pump if you want to keep costs down and if you have no pump space problem. If the pump is located in a pump house or in the basement the sound of its operation is seldom bothersome. If you are pinched for space or if absolute silence is more important than the somewhat higher cost, favor a submersible pump.

If you do the installation job yourself select a pump that can definitely be used with plastic pipe. Most home-sized jets, and all but the very large submersibles, can be used this way. But check with the dealer.

Use only the pipe sizes specified by the manufacturer, and be sure all connections are tight. Flexible plastic pipe should be pushed all the way onto connection fittings and tightly clamped *only* with clamps made for the purpose. On jet systems, one pipe is larger than the other. Be *sure* the correct size is connected to its matching fitting both at the pump and the jet. A separate length of plastic tail pipe leads downward from the jet to the foot valve. This reduces the length of double pipe required, but the tail pipe length should not exceed that recommended by the pump manufacturer.

The sound of pumps mounted above ground in basements or utility rooms can be reduced in several ways. The pump base, for example, can be bolted to the floor with rubber washers above and below the bolt holes, and rubber ferrules inside the holes. Use a slightly smaller bolt than the originals to make this possible, and use a metal washer above the upper rubber washer at each bolt hole. To prevent noise or vibration from being transmitted through the water supply pipes, connect the starting pipe to the supply tank with a 2' or 3' length of flexible plastic pipe.

If the tank is located where condensed moisture dripping from it will create a problem in warm weather, you can lick the problem either by placing a drip pan under the tank with a hose leading to a drain, or by using a fiberglass-insulated tank to eliminate the sweating.

When lowering plastic pipe into the well, either with a jet or a submersible pump on the end of it, allow as large a bending radius as possible for the pipe, and try not to let it scrape any rough edges at the top of the well casing. As shown in the diagram, the pipe is attached at the top to a "sanitary cap" that seals dirt out of the well. These caps are made in both one- and two-pipe forms to suit the type of pump.

Throughout any pump-system installation job follow the manufacturer's instructions to the letter. There can be considerable variation between brands. Be sure, too, that all priming directions are carried out before the pump is tested. After the pump is started, remember that a little time interval will pass before it can fill the pipe and begin driving water into your storage tank. When this happens you can hear it. After the pump shuts off, your tank is full, pressure is available, and you can forget all about it. Many pumps operate untouched for years.

TROUBLESHOOTING. If you hear your pump starting and stopping at very short intervals the tank may be "waterlogged," that is, it may not have an air cushion in the top. This can result from a loose fitting in the top of the tank or from a faulty air regulator. You can test the tank top with thick soapsuds to see if the remaining air bubbles out. In some cases you can see where water is leaking out. (No air left.) Tightening the fitting or plug usually cures the trouble. If the leak is large remove the plug, coat it with pipe compound or teflon tape, and replace it. If there's no evidence of a tank leak try giving the air regulator a sharp tap with a light tool like a pair of pliers. Sometimes this will free a stuck part. To re-establish some air space in the top of the tank you'll have to drain some water from the tank, with the pump switched off. Usually this is best done by disconnecting one end of the plastic pipe leading from the tank to the plumbing-supply pipes. This permits gravity draining, as there probably won't be pressure enough to carry the water to faucets at a higher level. The closer the disconnection is to the tank the better, so air can bubble into the tank as the water comes out. This may be a bit slow, but it works.

Since air will always dissolve in water, your tank has always to be drained of water at regular intervals. To lessen this chore you might consider installing a valve or a pair of valves right next to the tank so that drainage is fast and simple. To eliminate the drainage chore completely you can replace your existing tank with a modern diaphragm pressure tank. This tank has a rubber diaphragm separating the water from the air. In this way no air is ever lost and you never have to drain the tank.

In winter, if the pump is in a cold area like a pump house, you may sometimes find little or no water pressure in the morning after an exceptionally cold night. This seldom means a frozen pipe, however. But it may mean the slim tube to the pressure switch has frozen, preventing the switch from clicking on at low pressure. You can usually cure this trouble simply by holding a cigarette lighter flame under the copper tube for a minute or so, moving it along the tube to warm it evenly. As soon as it thaws, the pump will start.

If you hear your pump starting at fairly long intervals during the night, even though no water is being used it may mean a pinhole leak in the plastic pipe or a leaky foot valve. Neither of these troubles are common. If it's a leak in the pipe it's most likely to be above the water level where you can spot it by removing the sanitary well cap and looking down the well with a flashlight. If it's a foot valve, or if you think it is, wait a day or so. Sometimes new foot valves seal better after a little use.

Low water pressure at the faucets can be due to setting the pressure regulator at the pump too low. It may also be due to a valve that is partially closed and/or a loose pump belt.

Frequent fuse blowing can be caused by a pressure regulator set too high. The pump motor overloads trying to build up pressure.

5 | Common Plumbing Repairs

FORTUNATELY, most plumbing troubles can be repaired by the average homeowner without professional help. In some cases, however, it takes an experienced eye to judge the situation and decide on the best approach. Sometimes, for example, what seems like a minor leak can be turned into a major job by the wrong repair method. In the pages that follow you'll find detailed repair procedures and tips on avoiding some tricky common pitfalls. And, on the bigger jobs, you'll find advice on when you're likely to be better off if you leave the work to a plumber.

LEAKY FAUCET. This is one of the plumbing ailments most of us encounter sooner or later. If the faucet is a conventional one, the repair seldom requires more than ten or fifteen minutes to complete.

To get started, you need a box of assorted faucet washers and a monkey wrench or other adjustable wrench whose jaws open wide enough to grip the hexagonal cap through which the shaft passes to the faucet handle. If the faucet dribbles around this shaft you also need a small package of faucet packing.

First, turn off the water at the shutoff valve nearest the faucet.

To avoid scratching the chrome-plated hex cap, wrap it with stick-on tape. Almost any type will do. Then use the wrench to remove the cap, turning it counterclockwise. If the cap seems to jam during removal, you can usually free it by turning the faucet handle in the direction you turn it to turn the water *on*. (Twin faucets may or may not turn on in opposite directions.) When the cap is screwed off turn the faucet handle in the *on* direction until the shaft can be lifted out of the faucet body. (The large screw threads near the bottom of the shaft are the ones that open and close the valve when the faucet handle is turned.)

At the bottom of the shaft you'll see a small fibre or neoprene washer like the ones in your assortment. It's held inside a shallow metal rim by a brass screw through its center. It won't look exactly like your new washers because it will probably have a circular groove worn into it—causing the drip. Remove the screw and the old washer. A screwdriver is all you need here. You may have to use the screwdriver not only on the screw but also to pry the washer free. Long use sometimes spreads the washer so it sticks in its metal recess. In any event, it's easy to get it out.

If your new washers are separated as to hot- and cold-water use, be sure to use the right kind for the job. (When you buy your washers, get the kind that can be used on both hot- and cold-water faucets.) You may find that the nearest size washer to the old one is just a little too big for the recess on the end of the shaft. If so, you can trim it to size in a minute or two by simply drawing its edge across a piece of medium-grit sandpaper. Turn the washer a little after each stroke so as

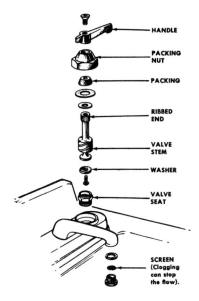

HANDLE

PACKING NUT

PACKING

RIBBED END

VALVE STEM

WASHER

VALVE SEAT

SCREEN (Clogging can stop the flow).

Parts of a Faucet

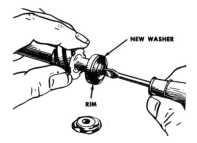

NEW WASHER

RIM

New washer is cure for dripping faucet. Simply remove center screw, pry out old washer, and replace with new one.

to trim it evenly all around. When you mount the new washer you can use the old screw. If new screws come with your washer assortment, however, as they sometimes do, you can use a new one and save the old one for an emergency. Turn the screw snugly, and be careful not to let the screwdriver slip and jab the hand that's holding the faucet shaft.

When the new washer is in place, replace the shaft in the faucet, turning it toward *off* to screw the big threads into place at the base of the shaft. Then push the hexagonal cap down the shaft and tighten it back on the faucet. That's all there is to the washer job.

You may find the faucet has a little different feel in use because the new washer is thicker than the old worn one. If the handle is the type that has to point in a certain direction for symmetry when turned off, you'll find that the handle can be removed from the shaft, usually by removing a chromed screw from the top. If it won't slide off easily, wrap a little tape around it and tap it upward gently with a pair of pliers. It doesn't call for hammer-banging. This type of handle is almost always "splined" on to the shaft. A number of fine teeth (sometimes like gear teeth) on the shaft fit into matching ones in the handle. Simply replace the handle on the shaft in the desired position. The teeth provide a wide position range. Then tighten the screw back in the top.

If the faucet dribbles around the shaft, cure the trouble while you have it apart for the washer job. You'll find a metal washer on the shaft just under the hexagonal cap. If you work this washer down the shaft, you'll see some fiber packing above it, inside the cap. (If there's a smaller washer above the first one, pull it down too.) Wrap just one turn of your new faucet packing (it looks like shiny, black vermicelli) around the shaft up against the old packing. Then push the washers back and you're ready to replace the cap on the faucet. As you tighten it down, the washers squash the new packing in around the shaft and stop the dribble. Details of the packing arrangement vary somewhat, but you can see how they work at a glance. Then adapt your re-packing to the particular faucet.

Faucets encountered in old homes are often a little different from the more up-to-date forms, but they usually take one of the standard washer sizes. And you can usually see how to disassemble them. One fairly common type found on some of the china wash basins of a generation ago (and highly prized today) often seems baffling at a glance. But it reveals its secrets quickly. It usually has a china faucet handle with four projecting members. A small nickel-plated hex rim holds the hot and cold labels (china disks) atop the center of the china handles. Once you remove the little hex rim by turning it counterclockwise, the job is under way. A screw under the china label disk is next removed to free the handle from the shaft. Then the flat, knurled metal nut (around the shaft) below is turned counterclockwise to free the little dome-shaped china covers that conceal the metal faucets. Just slide the nuts and the china covers up and off the shaft.

Usually you'll find two hexagonal caps (one inside the other) on the metal faucet. The upper one tightens the packing around the shaft. The lower (usually larger) one removes the shaft and washer unit when unscrewed counterclockwise. If the faucets have not been disassembled for a long time, however, it's a good idea to douse the threads of the caps with a penetrating lubricant like Liquid Wrench before you try taking anything apart. If a lot of muscle power is required on the wrench, repeat the penetrating lubricant treatment for several days. (You can put the handles back on the faucets, but leave the china covers off —so you can use the faucets but lubricate them, too.)

The reason for all this preparation: too much wrench power on these thin-shelled brass parts can actually break them. And in most cases replacement parts can't be found. If the unit is the typical form with a central spout and integral drain stopper mechanism, it's one of the highest-priced faucet assemblies on ths list—so don't take needless risks when you repair it.

Once the hex cap is removed and the shaft lifted out, washer replacement is the same as with newer types. To eliminate dribble around the shaft, just un-screw the inner hex cap, put a turn of packing around the shaft, and tighten the parts back together.

O-ring faucets. These differ from faucets that use fiber packing to seal their valve stems in that one or two O-rings, small rings of rubber, are used instead. The rings fit into annular grooves on the stem and contact the inside of a short tube that is held in place by a cap nut. The O-rings last many years, but even-tually must be replaced. The only point to bear in mind when doing so is that the new rings must be exact-size replacements. If the ring is too large or too small by a hair's breadth the stem will leak.

Diaphragm faucets. These utilize small, hat-shaped rubber diaphragms that act as both washer and O-ring. Sometimes the stem will carry an O-ring behind the diaphragm. Diaphragm faucets operate without leaking even longer than

O-ring faucets because the little "hat" is free to turn on the faucet stem. This eliminates the grinding action that results when a standard, fixed washer is turned against its seat. When the diaphragm eventually wears out, it is a simple matter to pull it off and press the replacement on the end of the stem.

Single-lever faucets. The task of mixing hot and cold water is simplified by this type of faucet. Operating the lever enables the user to select hot, cold, or a mixture, at the desired flow.

There are two designs in common use. One can be termed a slide valve and the other a ball valve.

The slide valve arrangement utilizes a fixed brass disc with three holes leading to the hot- and cold-water supply pipes and the discharge spout. Atop this disc is a second disc, sometimes made of brass, sometimes of porcelain and sometimes of plastic. The second disc has three corresponding holes, each ringed with rubber grommets. The holes are interconnected. The upper disc is coupled to the control lever and slides over the lower disc. Faucet operation depends on the relation of the upper three holes to the lower holes.

Wear occurs as the upper disc is slid back and forth. When the faucet drips it is time to replace the grommets. This is done by disassembling the faucet and simply pulling the old grommets out and inserting the new.

The ball valve differs in that the moving portion of the valve, the portion attached to the control lever is shaped like a ball. Like the slide valve, the ball is also pierced by three holes, and also like the slide valve design, the relative position of the holes in the ball to the balance of the faucet determines flow. However, instead of a flat lower disc or section, the ball valve faucet has two or three spring-supported grommets which lead to the feed pipes and spouts. In addition, the ball is held in place and sealed against water leakage by a circular packing ring atop the ball. In some designs there is also a pair of O-rings between the rotating spout and the body of the faucet.

When ball-valve faucets leak around their stems (under the lever) the compression nut—the large nut (ring) under the lever—may be loose. Try tightening it. Or the packing may be worn. In some models the aforementioned pair of O-rings may be worn. Shut off the water, disassemble and inspect.

When a ball valve faucet drips, one or more of the grommets are worn. Disassemble and replace. You can purchase a replacement kit of parts for most of these faucets. They include grommet springs which should be replaced along with the grommets.

CLOGGED DRAINS are often a more urgent problem. If the basin or sink is nearly full, bail out some with a pot and dispose of the water down an unclogged drain. This prevents a lot of floor splashing. But leave enough water to cover the rubber part of the force cup, the first tool to try.

To get the best results with a force cup, tip it to spill the air out as you place it under water and on to the drain opening. This sends a more solid punch along the pipe. If you're working on a double sink, have someone hold the stopper in the other sink firmly. Otherwise your first stroke with the force cup will pop the stopper out and possibly shoot a geyser of dirty water up to the ceiling. (Naturally

REPAIRING A SINGLE-LEVER FAUCET

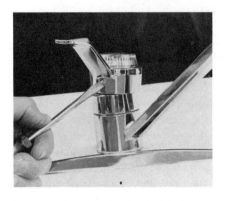

1. Loosen the setscrew under the handle.

2. Remove the handle by pulling it up.

3. Unscrew the cap.

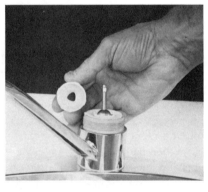

4. Remove the plastic disc that lies under the cap.

5. Lift the slide mechanism up and out. Note large O-ring that seals mechanism to faucet housing.

6. The two grommets in the bottom of the housing act as washers. Replace them if faucet leaks.

Inner parts of a ball-type single-lever faucet. Ball rotates beneath the circular packing ring (right). Spring-supported grommets and springs are shown at left.

the stopper must be removed from the drain on which the force cup is being used.) If you're working on a wash basin with an overflow opening have someone hold a sponge over the opening, for the same reason. These drain openings are often so angled as to shoot the dirty geyser at your midriff.

As you work with a force cup try to sense the rhythm of the water column you are sending back and forth along the pipe. If you catch the timing right, you can get more surge power with less effort. At intervals of about a quarter minute, pull the cup free of the drain forcefully. This will bring some grimy sludge into the sink, but it helps dislodge the blockage. Watch for signs of a sustained whirlpool that indicates the drain is cleared. Once this occurs, let the sink drain completely, but put the strainer in it to prevent the grime and solid matter from getting back into the pipe. You can scoop out the mess with paper napkins and flush it down the toilet after the sink empties. (Not too many paper napkins down the toilet at one time.) Then heat a big pot of water to boiling and pour it down the drain to clear any remaining grease. Don't use boiling water if you have plastic drain pipes.

If a force cup doesn't clear a clogged drain in four or five minutes of working time, it probably can't do the job at all. So quit and turn to the tool that can handle problems too tough for the force cup.

USING CHEMICALS TO OPEN DRAINS. Chemicals can only be used safely to open drains when there is some water movement. While chemicals — when they work — are a lot less messy and troublesome than force pumps or snakes, when they don't they leave a dangerous mixture of alkali or acid in the drain pipe. Then the problem of a plugged drain alone is compounded by the presence of a dangerous chemical.

Chemicals are fairly effective on all drains except toilets. There is so much standing water that you have to use several times the normal quantity. Even then, results are not always satisfactory because serious toilet stoppages are always caused by solid, nonsoluble objects which the chemicals will not attack to any appreciable degree.

Use any of the alkalis — lye, Drano, 99, etc., for kitchen sinks which are clogged mainly by grease. Use the acids — Clobber, Bust Loose, etc. — for lavatories, tubs and showers. The alkalis will cut grease by converting it into soap, but will not touch hair, which is a major culprit in lavatories, tubs and showers. 99 (a product name) is the best of the alkalis because it is pure sodium hydroxide and combines with grease to form a soft, highly soluble soap. Lye is caustic soda; it forms a hard soap. It is the cheapest but the least effective of the chemicals.

When using lye, wait for as much water as possible to drain out of the sink or

Clobber and similar liquid drain openers consisting of concentrated sulphuric acid are the most effective of all commercial drain-opening chemicals. Acid will attack hair as well as grease, which alkali will not. However, acid is very dangerous and should always be handled with extreme caution. Long-sleeved rubber gloves are a must.

tub. Do not simply pour the powder down the drain. It will solidify into an insoluble lump at the bottom. Instead, dissolve the powder in cool water, using a non-aluminum pot. Then pour the solution down the drain and give it plenty of time to work before turning on the tap.

Two very important precautions:

• Always use long-sleeved rubber gloves when working with the acid. It is concentrated sulphuric acid and very dangerous.

• Never follow an alkali application with acid or the other way round. Mixing acid with alkali will produce a minor explosion that will drive the combination out of the drain pipe and into the room.

The plumber's "snake" or clean-out auger is a flexible metal tool made in lengths up to 25'. Like a speedometer cable, it has a central steel spine wrapped tightly with finer steel wire. You push it into the clogged drain from the nearest convenient opening. Some of the shorter ones are so flexible they can be pushed

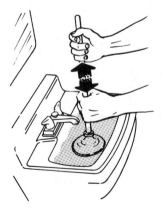

Suction cup works best when there is enough water in bowl to cover it. Tip the cup as it is immersed, to spill air out of it. Once on top of drain opening, it should not be tipped, but held as shown for tight seal during up-and-down pumping action.

WASHERS

Fixture traps that do not have clean-out plug can be removed by loosening large hex nuts at connection points. Don't lose rubber washers, as these seal connections when trap is in place. Trap usually has rubber-washered slip joint at fixture end, modified flare joint at drain pipe end. Both types are separated by simply unscrewing large hex nut.

right down the sink drain opening, through the U-trap under the sink, and on into the pipe. If you know your sink is close to the main stack (see Chapter 1) you can use this type. It will reach from the sink to the main stack, which is not likely to be clogged, so it can clear the stoppage. (If a main stack is clogged — a rare occurrence — all fixtures above the stoppage will fail to drain.)

If the drain pipe is a long one, you need a longer snake, which will be a little stiffer. In this event your first step is to bail out the water in the sink. Then place a bucket under the trap or gooseneck in the drain pipe under the sink, and remove the clean-out plug from the bottom of the trap, if the trap has such a plug. You need an adjustable wrench for this. After the remaining water has drained into the bucket, see if you can work your snake through the clean-out hole and into the drain pipe beyond. If you have to force it, don't. You'd only kink it. Instead, disconnect the trap from the sink and the drain pipe. If the plumbing is conventional, this is easier than it may seem at a glance. The big hex nuts that hold the chrome-plated trap sections together turn freely on one of the sections and thread on to the other. A rubber washer between the parts seals the connection. To separate the sections turn the big hex nuts counterclockwise in relation to the threads until they come free. Then slip the sections apart. This may call for a little strategic wiggling, but seldom more.

Using the snake is a simple matter regardless of the way it enters the pipe. The commonest types have a little tubular crank that slides over the snake itself. You can lock this in place anywhere along the snake, with a thumbscrew. This lets you rotate the snake as you push it farther into the pipe and is very helpful in getting it past snags and through stubborn blockages. Keep the crank tight on the snake just a few inches out from the point of entry into the pipe, and turn it slowly as you push. As the snake moves farther into the pipe loosen the crank, slide it back a few inches, and retighten it. You can usually feel the blockage easily when you hit it. If the snake becomes hard to turn, back it out an inch or so until it turns freely, then crank it into the blockage again. You can clear most blockages this way within a minute or two after the snake reaches them. When you can move the snake past the blockage freely, slide it back and forth a few times to make sure you've broken up the solid material. Then run hot water through the pipe as soon as you can to wash away the remnants.

For long, clogged sewer lines there are other flexible cleaning tools. These are sometimes called "sewer rods," though many think of them as plumber's

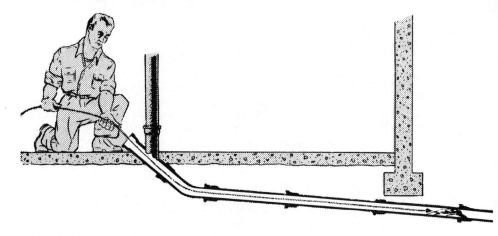

After draining as much liquid as possible from drainage system above, house drain clean-out plug may be removed to permit insertion of plumber's snake or clean-out tape. Small snake is used in the same way in small drain pipes after removing fixture trap.

"spring tape." They are made of flat, flexible steel tape coiled on a reeling frame. Like the smaller, more flexible types, they have an enlarged, pointed end to penetrate blockages.

BLOCKED SEWER LINES. The first problem in cleaning out a blocked sewer line is disposing of the sewage that has accumulated in the pipes above the clean-out plug where your tape will enter. It helps to drain all sinks and fixtures, and if the water in the toilet bowls is at the overflow level, bail them back to normal with a bucket. Then, before you loosen the clean-out plug in the basement, place as big a container under it as possible. Loosen the plug just enough to let the fluid in the pipe run into the container at a moderate rate. Then, as the container fills, tighten the clean-out plug to stop the flow. You can repeat this process until no more fluid comes out. Then you can safely remove the plug and go to work with your clean-out tape. This type doesn't have a sliding crank, but you can turn it by turning the lightweight metal rack on which it is coiled. You uncoil it from the reel as you push it into the pipe. Then, re-reel it as you pull it out after cleaning the stoppage. All snakes and tapes should be uncoiled again out of doors and thoroughly hosed down to remove dirt and bacteria. Then they can be recoiled for future use. In using the tape be careful not to let either end spring out and strike you.

CLOGGED TOILETS are cleared either with a two-way force cup or a closet auger. The force cup, which can also be used on other drains, has a fold-out section of the rubber cup. This fits snugly into the outflow passage of the toilet bowl. You use it like an ordinary force cup. The closet auger is really a short snake that extends from a hollow tube with a crank handle at the other end. The tube makes it easy to insert the snake portion into the outflow passage of the toilet. Then you crank the handle as you move the snake farther in until the

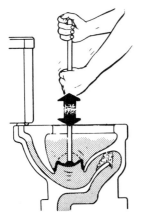

Dual-purpose force cup with wide, fold-out br m is used in toilet bowl to clear stoppage. Use it in same manner as ordinary force cup.

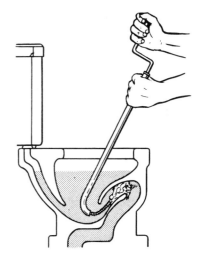

When force cup can't clear bowl stoppage, use "closet auger," a short, flexible plumber's snake with a tubular shaft and crank handle at the top. Turn crank as you work snake through passage toward and through stoppage.

stoppage is cleared. This is usually a quick job. Before you begin it, however, take a careful look at the outflow passage of the bowl. Sometimes, if there are small children in the house, you may find the trouble is caused by a doll or other toy that you can see and remove without further ado.

LEAKY PIPES are usually easy to fix unless the leak occurs at an old threaded fitting that has rusted or corroded badly. Even if the section of plumbing can be taken apart so the fitting can be replaced, old corroded pipe is likely to break altogether when you apply wrench force to it. If you haven't the time for the fairly extensive job that might then be necessary, you'll be wise to leave it all to the plumber.

The easiest way to stop a leak in a straight run of pipe is with a little ready-made piece of hardware often called a "pipe leak clamp." It's made in two halves with a rubber liner and a tightening bolt. Buy it sized to match the diameter of the leaking pipe. Then link the halves together around the pipe with the rubber liner over the leak, and tighten the bolt. Design details vary a little, but that's all there is to it. The repair is likely to last for years. Pinhole leaks of the type cured in this manner often occur along the underside of an old iron cold-water pipe. They often rust through over the years from the outside due to water droplets that collect from condensation. (If you have some rusty cold-water pipes that haven't developed any leaks give them a coat of rust-inhibiting paint.)

Leaks in threaded joints. These are usually caused by pipe that has loosened. When there is a union nearby in the line, you can back off a bit on the union and then tighten the leaking joint. Following, the nut on the union is tightened again. If you don't mind a little water, pressure can be left in the pipe.

When there is no union in the line, try slowly and carefully to tighten the joint anyway. Sometimes the following joint can be loosened a fraction of a turn without causing it to leak.

If tightening the joint doesn't help, use a pipe cement to seal the leak. The pipe must be drained of water and the leak area thoroughly dried, cleaned of rust and roughened with sandpaper. Apply the cement in a thick, circumferential layer and give it a day or so to dry hard.

If that doesn't do it, you have to take the leaking joint apart and correct any of the following possible causes of leakage: too much thread cut into the pipe; too little thread; poorly cut thread; lack of pipe dope; thread destroyed by rust.

Leaks in plastic pipe. These can be repaired with a patch made of similar plastic and solvent-welded in place. For example, use half of a slip fitting of the proper size, cut in half lengthwise. Drain the pipe. Dry and clean the area around the hole and "weld" the patch in place. If the hole is large, use a pair of pipe clamps as well as solvent to hold the patch in place.

Leaks at plastic pipe joints can on rare occasions be sealed by emptying the pipe and joint, thoroughly drying the two and then forcing solvent into the joint. If this doesn't work, you have to remove and replace the fitting. Unfortunately, this can only be done by cutting off the fitting along with a section of pipe. Afterward, you have to install a duplicate of the original fitting plus a length of pipe and a slip fitting. This is why it is important to be very careful when making plastic joints.

Leaks in copper pipe that occur at soldered (sweated) joints are seldom a problem. Cure the trouble as described for testing plumbing in Chapter 3. Drain the pipe of all water. Then heat the joint with a propane torch to the solder melting point. Flow solder into the part of the joint that leaks. And check all other parts to be sure they have ample solder. If other fittings are close, you can avoid softening their solder (through transmitted heat) by wrapping them with wet rags.

If a copper pipe is crushed or punctured, as can happen accidentally in the basement or garage, the repair is almost as simple. Shut off the water, as usual, and drain the pipe. Then use a tube cutter to cut out the damaged section. If the pipe has been crushed be sure to cut back far enough to reach *round* pipe. Then cut a new piece to the same length as the removed section. Place a "slip coupling" on each end of the new piece, and slide it inward an inch or so. This coupling is simply a short piece of machined tube that fits over the regular tube. It does not have the usual internal shoulder to prevent it from sliding along the regular tube. (The inside of the fittings should be cleaned before use with steel wool or very fine sandpaper.) Clean the cut ends of the original pipe and the ends of the replacement piece so that all joining ends are shiny. Then apply flux to all joining ends, and slide the slip fittings over the joints. You can mark the pipe with soft pencil so that meeting ends come inside the center of the coupling

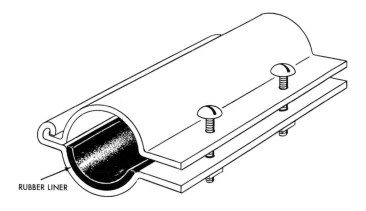

RUBBER LINER

Pipe-leak clamp, though varied in detail according to manufacturer, looks like this, Remove screws to open clamp so it can be placed around pipe at point of leak. Rubber liner should be directly on leak. Close clamp, replace screws, and tighten firmly to seal leak.

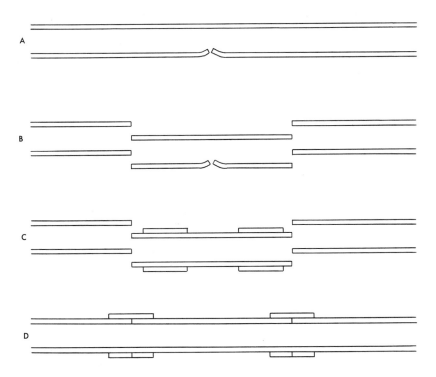

A

B

C

D

Repairing punctured or damaged copper plumbing tube. A. Damaged section of tube B. With pipe (water-supply or drain) empty, cut out damaged section either with tubing cutter or hacksaw, making both cuts far enough from damage to be on smooth part of tube. C. After reaming cut ends of tube where section has been removed, cut and ream section of new tube of a length matching removed section. Push "slip coupling" onto each end of new section. D. After cleaning all meeting ends of tube with steel wool or very fine abrasive, coat them with soldering flux, and slide slip couplings over joints. Finish by soldering the joints as described in Chapter 3.

fittings. Then use a torch to heat the joints, and flow solder into them. Be sure you apply solder to both ends of each coupling until the silvery ring all the way around tells you the connection is sealed.

Drain-pipe leaks are more easily fixed in a pinch because they carry almost no pressure. Small drains, as from sinks, can usually be cured of leaks, even at joints, by washing the area with detergent, drying thoroughly, then wrapping snugly with rubber electrician's tape. Overlap the tape edges. This makes a quick but long-lasting repair on any type of metal drain pipe. If you have time to use the methods just described for water pipe, however, do so.

Cast-iron drainage pipe. Leaks that occur at the leaded joints can usually be cured in short order by spreading the lead a little (as in making the joint) with a calking tool. If you lack the tool, you can make a substitute by flattening the end of a big nail with a hammer and filing a sharp, straight edge on the flattened end, like a chisel. Just tap the sharp edge into the lead at a number of points in the leak area. You don't have to cut deep. As the cuts spread the lead and wedge it against the pipe surfaces, the leak stops.

HOW A TOILET WORKS. Knowing how a toilet operates makes it much easier to diagnose and cure its troubles.

When you operate the flush handle an inner level attached to the handle shaft lifts a soft rubber "ball valve" (often called a stopper ball) out of a large opening in the bottom of the flush tank. This allows the entire contents of the flush tank to flow into the toilet bowl at a much more rapid rate than would be possible if the water were simply flowing from the water-supply pipe. (The purpose of the flush tank is to supply this sudden surge of water.) Once the rubber stopper ball is lifted out of the opening in the tank bottom it floats upward rapidly in the water remaining in the tank, as the stopper ball is hollow and filled with air. That's why you don't have to hold on to the flush handle to keep the flushing operation going. The stopper ball won't drop back into the opening until the water level in the tank falls to the level of the opening.

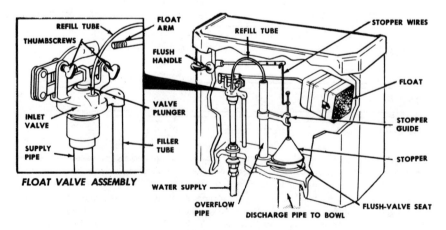

Parts of Toilet Flush Mechanism

As the water level falls, a large float also falls with it. A metal float arm reaches from the float to a "float valve" at the opposite side of the tank. This is really the water inlet valve that refills the tank. As the float falls with the water level it opens the float valve and turns on the water to fill the tank again. But there's no inlet–outlet conflict because the water comes in through the float valve much more slowly than it goes out through the float valve. (That's why it takes a toilet tank much longer to fill than to flush.)

When the tank is empty and the stopper ball has dropped back into the flush valve opening at the bottom of the tank, closing it, the float valve starts filling the tank. At the same time it squirts a little stream of water through the "refill tube" into the overflow pipe in the tank. This overflow pipe leads into the toilet bowl. Thus, if something goes wrong with the tank mechanism so that water continues to flow into the tank, it simply flows into the toilet bowl, then on into the soil pipe and sewer. The purpose of the refill pipe, however, is to make certain that the toilet bowl is completely filled again after flushing. (The flushing process siphons much of the water out of the bowl, so it must be refilled in order to maintain a water seal against sewer gas, like a trap, as described in Chapter 1.)

As the water rises in the tank, the float rises with it and shuts off the float valve. This also stops the water flow through the refill pipe. The stopper ball doesn't float up out of the flush-valve opening because it is designed so that the rising water exerts a downward pressure on it—and there is no water under it— since it has sealed the large pipe below it. This pipe (called the discharge pipe or "spud") is empty after the flushing is completed.

"Singing" toilets. One of the commonest troubles is the "singing" toilet that doesn't shut off after flushing. This often happens because the stopper ball fails to seat properly in the flush-valve opening. It may get off-center or become misaligned through tilting. If this is the problem, you can see it easily by removing the top of the tank. Water will be flowing into the tank through the float valve because the float is at the bottom of the tank, but the tank won't be filling because the stopper ball isn't sealing the flush valve opening. Often you can cure the trouble simply by jiggling the flush handle to bounce the stopper ball into its proper position. But if the trouble recurs, some adjustment or part replacement may be necessary.

To check on the flush valve, shut off the water to the toilet, flush it, and watch the operation through the open top of the tank. The tank won't refill with the water shut off so you can repeat the flush handle operation several times to see just how the stopper ball fails. Sometimes the stopper guide manages to swivel around the overflow pipe (which supports it) enough to get the ball out of line with the flush-valve seat. If so, loosen the setscrew that hold the guide to the overflow pipe, turn the guide so the stopper ball seats as it should, and retighten the screw. You need a short screwdriver for this because of the cramped quarters inside the tank. Some guides also have a length adjustment that lets you vary the distance outward from the overflow pipe to the guide rod that attaches to the stopper ball. If this has loosened, the stopper ball may be landing on the inner or outer edge of the valve seat instead of contacting it all the way around. Whatever the misalignment, it's easy to spot with a few trials of the flush handle.

If water continues to run out through overflow pipe when float is all the way up (and tank full) you can usually cure the trouble by bending the float arm to lower float. If float has a leak, and remains at bottom of tank when tank is full, unscrew old float and screw new one on to threaded end of float arm.

Stopper ball can be removed and replaced with new one without removing stopper wires from guide or flush handle arm. Just hold stopper ball still with one hand while unscrewing lower stopper wire with the other. Then screw lower stopper wire into replacement stopper ball in same manner.

If the stopper ball is seating properly but still not stopping the outflow of water, the trouble may lie either in the ball itself or in the valve seat. If the valve seat looks pitted, smooth it with steel wool or very fine wet-or-dry abrasive paper. (This type of trouble usually allows the tank to fill, but keeps it singing.) If the stopper ball itself looks lumpy, misshapen, or dimpled, replace it. It simply unscrews from the thin brass rod, sometimes called a stopper wire, that lifts it.

If the toilet will flush completely only if you hold the flush handle down during the entire flushing process, the stopper guide is mounted too low on the overflow pipe. This prevents the stopper ball from rising high enough to clear itself from the outrushing water. So, unless held up, it is sucked back into the flush-valve opening. Simply raise the stopper guide a little, tighten it in place, and try flushing again. If the trouble persists, raise it a little more until the flushing process runs its full course without need for holding the handle. Usually raising the guide about half an inch does the trick.

If the tank fills but the toilet keeps on singing and water flows out through the overflow pipe, you can easily spot the condition by lifting the lid off the flush tank. This kind of trouble is caused by failure of the float valve to shut off the water after the tank has filled. A leaking float can cause it, though this is rare. But if the float is at the bottom of the tank, that's the cause. Shut off the water to the toilet, flush it, and unscrew the old float from the threaded end of the float arm. A new float from the hardware store will screw on to the same threads.

If the float is on the surface, the trouble can often be cured by simply bending the float arm downward slightly so as to push the float a little deeper in the water. This way it closes the float valve with the water at a lower level. Usually this shuts off the water before it reaches the overflow. If it continues to flow in

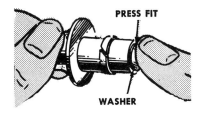

If float valve fails to shut off when float is up, the trouble is usually in the washer at the base of the valve "plunger," the small piston that is moved up and down by the levers at the top of the valve. When the thumbscrews and levers are removed, you can lift out the plunger and replace washer.

spite of this float-arm adjustment, the chances are the valve itself needs repair. This most often means replacing the rubber washer on the bottom of the valve plunger. Shut off the water, remove the little thumbscrews (see valve-assembly drawing) and lift out the valve levers that pivot on the screws, along with the valve plunger they operate. This is usually easiest if you unscrew the float arm from the valve first. You'll see a rubber washer at the bottom of the plunger. Sometimes it is held in place by a screw through the center, like a faucet washer, or it may be held simply by its tight press fit inside a metal rim. Although these washers are standardized to a considerable extent, it's a good idea, if convenient, to take the old washer or the complete plunger with you when you buy the new washer.

Important: In all work on float valves or on flush mechanism parts that are mounted on the overflow pipe, avoid unnecessary force or pressure in any direction that would tend to bend the valve supply pipe or the overflow pipe. Both are relatively thin-walled and may break at the bottom where they are threaded into fittings in the bottom of the tank. (The threads cut part way through the thin metal walls of these pipes, resulting in a weak point.) The same caution applies to the refill tube. This must sometimes be bent out of the way, as when removing the stopper guide completely.

The new flapper valve type of flush valve is often used to eliminate stopper ball and lift wire and guide troubles. The flapper valve is an all-rubber unit that is slipped down on the overflow pipe after removing the stopper guide, the old stopper ball, and all its lift wires. A length of small stainless-steel chain links the flapper valve to the regular flushing arm inside the tank. The all-rubber flapper mechanism is a nearly foolproof one-piece unit held in place by a friction-grip rubber ring around the overflow pipe. Complete installation directions are printed on the container in which you buy the unit.

Before buying one, however, carefully measure the distance outward from the surface of the overflow pipe to the center of the flush valve opening. Do this with water turned off, flush tank empty. In most cases this will match the distance from the friction-grip ring of the flapper valve to the center of the flapper valve stopper ball. In some cases, however, the distances don't match, so the flapper valve cannot be used, as it would fail to seat properly.

To replace a float valve, first shut off the water to the tank. (Complete valve replacement may be required if you can't buy parts for an old or discontinued valve.) The water-supply pipe to the tank is connected to the valve pipe inside the tank by means of a rubber slip joint and nut. The base of the valve pipe is sealed into the tank by a rubber gasket that is squeezed down snugly against the inside of the tank bottom by a metal nut on the outside. After shutting off the water and flushing the tank to empty it, unscrew the slip-joint nut by turning it counterclockwise. A large adjustable wrench is good for this. When it's free from the threaded base of the valve pipe, remove the nut that holds the valve pipe in the tank, and you're ready to lift out the complete valve unit. It usually helps to use a small block of wood wedged against the inside hex nut on the base of the valve pipe to prevent the pipe from turning when you use a wrench on the outside. Or you can tighten a second wrench on this inner hex nut and let its handle come snug against the inside surface of the tank. But in all this work use care not to crack the tank. If a wrench is likely to swing against the tank, wrap the wrench handle in a few layers of cloth.

Installing the new valve is simply a matter of reversing the steps followed in removing the old one. It's usually wise to replace the rubber gasket at the same time, also the rubber slip-joint washers. Be sure the new valve is positioned, when tightened, so that the float arm will extend in the right direction.

Diaphragm-type float valves have been recently introduced as replacements for the plunger-type float valves. Both are installed and removed exactly the same way, and both are operated by the rise and fall of the float. The difference lies in the valve mechanism. Whereas the plunger type utilizes a long plunger with a washer at its bottom to seal the water inlet, the diaphragm design utilizes a diaphragm alone or a diaphragm plus a short plunger and washer. In all models

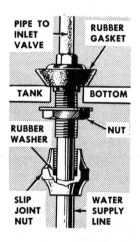

How float valve is mounted in tank. If you must replace entire float valve, first unscrew slip-joint nut all the way, from threads protruding from tank bottom. This disconnects water line — so water to toilet must be turned off in advance. Then remove large nut directly above slip-joint nut. Valve pipe and valve may now be lifted out of tank. Mount replacement valve by simply reversing the steps. It usually comes with a new rubber gasket on it. If not, buy one to fit it. Don't re-use old gasket.

If you have a leak between toilet bowl and floor, trouble is almost always at putty seal between bowl and closet flange of waste pipe. Unbolt bowl from floor, turn it upside down, and remove all traces of old putty and plaster. Then replace the flange-to-bowl seal with a modern wax or rubber ring-gasket from plumbing supplier.

Working end of a diaphragm-float valve. Float swings on arm that acts directly on push rod, which in turn acts directly on valve inside. Tank fills more rapidly with this type of valve; shutoff is more abrupt, too.

The inside of a diaphragm-float valve. Central washer can be pried out and replaced when necessary. It actually seals top of feed pipe. Second washer or diaphragm provides additional sealing at shutoff.

Close-up of a Speedee fitting made for connecting float valve. Hat-shaped plastic washer provides the seal between the fitting and the end of valve. Flange nut holds the fitting.

the diaphragm, or the diaphragm, plunger and washer are moved up and down by means of a short rod that is linked directly to the float arm. There is no intermediate linkage. The result is that the tank is filled more rapidly and shut-off occurs more quickly.

To service a diaphragm float valve, remove the upper half supporting the float. In some designs two machined screws are backed off. In others, the upper half unscrews. Defective washers and diaphragm can then be replaced.

The water-supply is connected to the newer type of float valves by means of a Speedee fitting, a short copper tube terminating in an expanded end which seats a cup-shaped, plastic washer. A flange nut positioned behind the cup is used to pull the fitting into place and hold it there. Some of the more recently manufactured plunger-type float valves are also constructed to accept Speedee fittings.

Leaks around the toilet base at the floor connection are easy to fix in some cases, difficult in others, depending on the type of unit. The problem, itself, is most likely to occur with old toilets that were installed before present day sealing materials were available.

If the flush tank is mounted directly on the bowl, it will be necessary to remove the tank from the wall. Look inside the tank near the top for hanger bolts through the back of the tank into the wall. Look also for bolts holding the tank down on the bowl. These will show, if present, at the bottom of the tank, close to or through the flush-valve assembly. To remove the tank, shut off the water to it, flush it, remove all mounting bolts, and lift it from the bowl. Many modern types are not bolted to the wall.

If the tank is not mounted directly on the bowl, but connected to it by a short piece of large, chromed pipe (called a spud), it may be possible to remove the bowl from the floor without demounting the tank — especially if the spud is an elbow or S-curve type. The spud is connected at both tank and bowl ends by a modified rubber-gasket-slip-joint arrangement that requires a special wide-jaw wrench called a spud wrench. (This is handy for other jobs requiring wide jaw gaps.) By loosening the spud nuts and slipping them free, onto the spud pipe, it is frequently possible to wiggle the spud completely free, and remove it. This permits you to remove the bowl from the floor without removing the tank from the wall.

To remove the bowl, first look for the hold-down bolts on the bowl base. These are often capped with plastic or ceramic covers (about olive size) to improve appearance and reduce corrosion. Most types pry off easily. Others unscrew. Try unscrewing first. Then pry. But protect the base of the bowl.

Once the hold-down bolts are removed the bowl can be lifted free. (It's wise to bail out as much water from the bowl as possible before this.) If the bowl

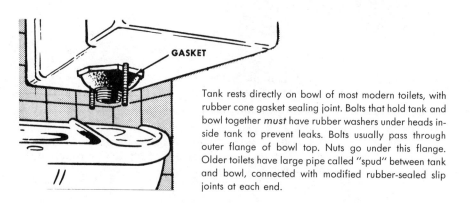

GASKET

Tank rests directly on bowl of most modern toilets, with rubber cone gasket sealing joint. Bolts that hold tank and bowl together *must* have rubber washers under heads inside tank to prevent leaks. Bolts usually pass through outer flange of bowl top. Nuts go under this flange. Older toilets have large pipe called "spud" between tank and bowl, connected with modified rubber-sealed slip joints at each end.

sticks, try tipping it slightly to one side to free it. When it's free, lift it straight off the bolts, drain the remaining water into a bucket, and turn the bowl upside down. You may find it was originally sealed to the flange that connects it to the soil pipe by anything from plain putty to plaster. You can use a sharp screwdriver, beer opener, and icepick to scrape out the remains of the old, leaky sealing material. Clean the floor flange, too. When the old sealant has been completely removed from both parts, use a modern, ready-made toilet floor-flange seal. This is usually applied to the floor flange so the bowl base can be pressed down into it. But some types are applied to the base of the bowl while it is inverted. Just follow the directions that come with the type you buy. The soft wax type is usually both simple and effective, and likely to last indefinitely.

Once the new seal is in place and the bowl reset on the hold-down bolts, complete the job by reversing the steps that started it. If there is any chance that the tank flush-valve seat needs replacing, this is a good time to do it—before the spud is replaced. You may also putty around the toilet bowl base for neat appearance.

In any toilet repair work it's a good idea to have a look at the actual type of parts you'll be working on—before you start the job. One of the best places to see the important details is at a housewrecker's storage yard. Here, you're likely to see both tanks and bowls, with and without spuds and other parts, and you're likely to find partially disassembled units that give you a clear idea of the details.

6 | How to Alter or Extend Your Plumbing

As YOU improve your home, the plumbing changes you may need can range all the way from adding another outside faucet to installing a lavatory or a complete bathroom. Whether it's wise to do part of the job or all of the job yourself depends on the individual case. The details that follow will serve as a guide in deciding how much to do and how to do it.

EXTENDING THE WATER-SUPPLY SYSTEM. This is usually an easy job for the homeowner. If the water-supply system is of copper tubing, the addition can often be completed in a matter of hours.

The first step, of course is planning. This begins with the selection of a starting point for the new piping. In many instances you can remove an elbow fitting and replace it with a T fitting to provide the extra connection that starts the new run. If there's a considerable length of straight pipe leading to the elbow in one or both directions there will usually be enough flexibility to permit slipping the elbow off after the connection is torch-heated to soften the solder. After that, it's a simple matter to clean the inside of the T fitting and slip it on in place of the elbow. To make sure the new fitting will slip on where the old one came off it's a good idea to wipe the hot solder off the tube ends with a rag. This eliminates cold solder bumps that might prevent the new fitting from slipping into place. Before mounting the new fitting, coat the pipe ends with flux, as in new work. Then heat the joint and, when it reaches solder-melting temperature, flow in the solder to seal each connection.

The first section of the new piping should be connected to the fitting at this time. If you have to do the job in stages because of limited spare time you can solder a cap on the end of the new run's first section (or any other point) so you can turn on the water and use your regular system until you have time to finish the addition.

If it isn't convenient to flex an existing joint apart to start your extension, you can remove a short section of straight pipe, as described in Chapter 5 under Leaks in Copper Pipe, and replace it with another section that includes a T fitting for the new branch. However you start your addition, you can continue it either with rigid tubing or flexible tubing. If it is to be led through walls or carried across ceilings or enclosed floors, soft tubing makes the work much simpler and reduces the chance of leaks. (All this if your local code doesn't restrict your choice.)

If you prefer to use the rigid tube for neat appearance where plumbing is exposed, you can switch to the soft tube for the unexposed runs. Remember, too, that you can place a valve somewhere along the line at an accessible point and keep it closed while you're working on subsequent stages of the job. This elimi-

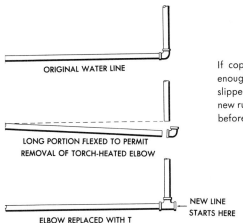

ORIGINAL WATER LINE

LONG PORTION FLEXED TO PERMIT
REMOVAL OF TORCH-HEATED ELBOW

ELBOW REPLACED WITH T

NEW LINE
STARTS HERE

If copper water tube leading to elbow is long enough to flex, so elbow can be torch-heated and slipped off, simply replace elbow with T to start new run of tube. Water must be drained from tube before you start the job.

nates the trouble of removing temporary caps as you go along, and it may come in handy for minor future repairs. Be sure you use a gate valve for this purpose, however, as this type, when open, permits full pipe diameter. The globe valve is the type found in faucets. Even when fully opened, its internal water-passage area is smaller than most residential water pipe, so it retards the flow to some extent beyond it.

There is a solderless method of connecting tubing to a copper tube by using a clamp-on saddle tap. This is a combination clamp and valve that is clamped firmly onto the tube to be tapped. With some saddles a hole is drilled into the tubing through a hole in the saddle and then the balance of the device is assembled. In others, no hole is drilled. The tube to be added is connected to the saddle's valve by means of a metal-ring compression joint. Then the handle atop the valve is screwed down. This punctures the water-filled tube and water flows

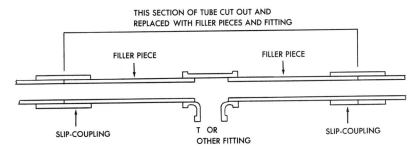

THIS SECTION OF TUBE CUT OUT AND
REPLACED WITH FILLER PIECES AND FITTING

FILLER PIECE   FILLER PIECE

SLIP-COUPLING   T OR
                    OTHER FITTING   SLIP-COUPLING

Where water-supply tube can't be flexed, you can connect into any straight run this way. First cut out a short section of the tube. Then replace this section with T and short filler lengths of tube to make up same total length as cut-out section. Connect this new assembly to existing tube ends with slip couplings. These couplings are slipped all the way on to the filler pieces so the T assembly can be set in place. Then they are moved out over the connecting points so the tube ends meet inside the midpoints of the slip couplings. All cut ends of tube should be reamed inside, cleaned, and fluxed. Joints are then soldered.

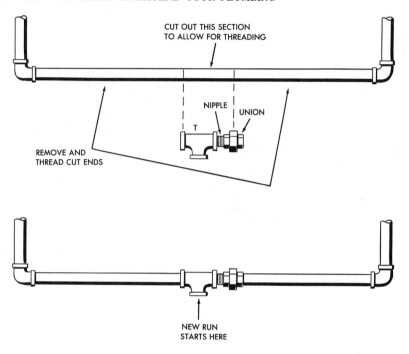

CUT OUT THIS SECTION
TO ALLOW FOR THREADING

NIPPLE   UNION

T

REMOVE AND
THREAD CUT ENDS

NEW RUN
STARTS HERE

To connect into run of threaded pipe (such as steel) cut out a section long enough to be replaced with a T, a nipple, and a union, assembled as shown. Then unscrew the remaining sections of pipe on each side of the cut-out section, and have their cut ends threaded. Screw the T on one section and replace it in the system. Screw the nipple into the T. Screw one end of the union onto the nipple. Screw the other end onto the other section of pipe, and replace it in the system. Then screw on the center hex of the union to seal the two halves of the union together.

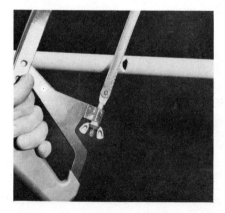

To tap into a plastic pipe, start by drilling or cutting a hole in the side of the pipe. Clean the edges of the hole with a penknife or file.

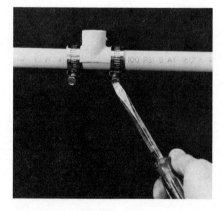

Cut a plastic T in half, lengthwise. Solvent-weld the T over the hole in the plastic pipe. Tighten two strap clamps over the ends of the T and the pipe. Now you can connect a plastic pipe to the T.

To connect a clamp-on saddle tap to a copper tube, start by positioning the two halves over the tube and tightening the bolts. Do not flatten the tube by overtightening.

Connect the second tube to the fitting by means of the metal-ring compression joint integral with the fitting.

Screw the handle down into the fitting. This pierces the tubing below and permits water to flow out of the tap. With this tap-on design it is unnecessary to drain the water from the tube.

into the new tube. With the latter type of saddle it isn't necessary to shut off the water.

To connect to a plastic water pipe you can follow the suggestions given for connecting to rigid copper tubing, or you can solvent-weld half of a plastic T to the side of the plastic water pipe.

Drain the water system and cut a hole into the side of the pipeline. Dry the area around the hole and rub with fine sandpaper. Then, with a hacksaw, cut a plastic T, of the proper size, in half lengthwise. Remove the burrs and sand the inside of the T. Now the T can be solvent-welded onto the pipe. This is done by applying a thin, even coat of the proper solvent (to match the type of plastic used for the pipe and T) to the outside of the pipe and the inside of the T. Give this coat a minute or so to dry, then apply a thick, even coat of the same solvent.

Press the T against the side of the pipe, directly over the hole, and hold it firmly in place for a full minute or more. Then put two pipe clamps over the arms of the T to reinforce the joint. You can now connect your new water line to the T.

If your new plumbing will lead to a new room or to an expansion attic, install your pipes before the inside walls are put up. This is a lot easier than stringing it through afterward. Any connections that cannot be reached later on without cutting into the wall should be soldered. These are likely to remain tight for the life of the house, though flare and compression connections may sometimes loosen slightly from vibration. The solderless connections, however, are very handy where you must work in cramped quarters close to flammable building materials. And, if you're using soft tubing the chances are you'll need fittings only at the beginning and end of each pipe run. If at all possible, limit your soft tubing to vertical or near vertical runs, or to lateral runs where enough pitch can be provided to clear all water from the wavy pipe when it must be drained to prevent freezing.

Adding an extra outdoor faucet is one of the simplest expansion jobs. Make your starting connection as described earlier. Use either rigid copper tube or galvanized pipe (whichever matches the rest of your plumbing) as it is necessary to slant the pipe downward toward the outside so it can be drained.

To keep the outdoor faucet from freezing in winter you can follow either of two methods. You can use a stop and waste valve in the pipe inside the wall and an ordinary sill cock outside. This is your only choice if, for some structural reason, the pipe to the outside faucet must be level or slanted upward toward the outside. The stop and waste valve should be mounted in the pipe so that the knurled waste cap is toward the outside (non-pressure) side. When the stop valve (handle-controlled) is closed and the waste cap removed, the water will drain from the pipe no matter which way it slants if the outside faucet is open. If it slants downward toward the inside, the water in the outer pipe will drain through the open waste cap into the cellar. There's not much water in the short pipe through the wall, so you can catch it all in a tin can. If the pipe slants downward toward the outside, the water runs out of the faucet while air enters through the waste opening. Be sure you don't lose the little cap, however; put it back as soon as the draining is completed.

Where you have a good downward pitch toward the outside you can use a freeze-proof wall faucet, which has the faucet handle on the outside of the house although the valve itself is actually inside the basement. A long shaft extends from the handle through the inside of the integral length of faucet pipe, to the inside. The design is such that the entire faucet and through-the-wall pipe section drains automatically every time you shut off the faucet. This also lets you run water from the outside faucet in winter any time you want it. It won't freeze while it's running, and the whole unit empties when you shut if off.

Adding a garage faucet. Even if your garage is attached to the house and heated, you'll be wise to take the same precautions against freezing as for the outdoor faucet—just in case somebody forgets to close the garage doors on a frigid day. If the garage is not attached to the house and is not heated, two more plumbing features are necessary. First, the pipe from the house to the garage

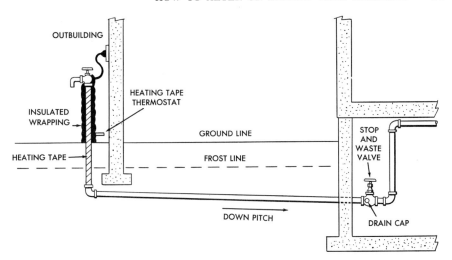

Water pipe for year-round use in unheated outbuilding should be protected with electrical heating tape from below frost line to highest point above frost line. Faucet must also be wrapped with tape. All above-ground sections should be wrapped with insulating tape. Use thermostatically controlled tape, and place thermostat above ground.

must be pitched downward toward the house and must have a stop and waste valve inside the basement. This is true even though the pipe is run below the frost line, underground. And, if the faucet is to be used in winter it must be wrapped with thermostatically controlled electric heating tape made for the purpose. The tape covers not only the faucet body, but the vertical section of pipe leading to it. The wrapping must extend below the frost line. Complete wrapping instructions will accompany the tape you buy.

Where a large number of threaded pipe connections are added, it's always wise to include a union or two (Chapter 3) in case the added plumbing should ever have to be disconnected for repairs. Unions are also available for copper plumbing. They make it easy to remove sections of plumbing or to disconnect appliances like water heaters without unsoldering the joints. In soft tube work, flare joints do the same thing.

ADDING DRAINAGE PLUMBING is relatively easy with copper plumbing, not so easy with cast-iron or galvanized steel. Before deciding whether to tackle this kind of work yourself, take a careful look at your plumbing. And take a careful look at your local plumbing code, if there is one. You can find out where to get a copy by calling the local building inspector's office. Some codes require special features not required by others. More about this as we go along.

To connect your new drain pipe you will either have to add a connecting fitting to your existing soil stack or to a secondary, smaller waste stack. In some cases you may have to add a new soil or waste stack. And, if the fixture is some distance from the stack, you'll probably need a "re-vent" pipe from the fixture's drain pipe to the vent stack above it.

Adding an extra sink is a good example of the basic procedure. When an enclosed breezeway is converted to a potting room for the home gardener, the extra sink often goes there. Photo hobbyists, too, need a sink in the darkroom.

The first step is charting the path the new drain pipe can take from the sink to the soil stack or waste stack. Depending on the situation, it may run from the fixture trap (directly under the fixture) along the wall under a counter, then down inside a wall to the basement, or it may run from the trap straight down through the floor and across the basement ceiling to the stack. There are countless other courses it may have to take, instead, depending on the individual case.

Where the drain runs from one level of the house to another, as from the first floor to the basement, it's best to use vertical, rather than slanted piping. (Short slanted sections may be used to get around obstructions such as parts of the house framing.) Lateral runs, as under counters or across the basement ceiling, should have a downward pitch toward the stack connection of from ¼″ to ½″ for each foot of run. The connecting fittings that join your vertical runs to your lateral runs provide for this pitch. Unlike water-pipe fittings, drain-pipe fittings are not always exactly what they seem. A "right-angle" elbow, for example, is likely to form about a 92-degree angle instead of the true 90 degrees. The extra 2 degrees aren't apparent to the eye when you look at the fitting, but that's all it takes to provide the needed pitch. The same applies to T's that are used to connect a vertical vent line at some point along a lateral drain line. If the T isn't marked for the direction of flow, you can tell which way the branch slants by temporarily inserting lengths of pipe into the fitting. A few feet of pipe reveal angles too small to discern in the fitting.

Adjust your plan so your new drain pipe enters the stack at a convenient location for a connecting fitting by varying the length of the vertical runs. Usually, it's possible to plan roughly with taut string, push pins, and nails. The chart in Chapter 3 shows how much extra length the various fittings add. For

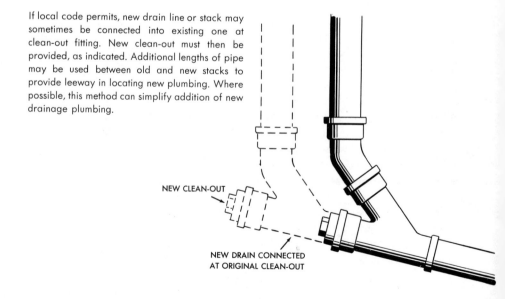

If local code permits, new drain line or stack may sometimes be connected into existing one at clean-out fitting. New clean-out must then be provided, as indicated. Additional lengths of pipe may be used between old and new stacks to provide leeway in locating new plumbing. Where possible, this method can simplify addition of new drainage plumbing.

NEW CLEAN-OUT

NEW DRAIN CONNECTED
AT ORIGINAL CLEAN-OUT

The easy way to tap into the middle of a cast-iron soil stack is to cut out a section of stack and replace it with a No-hub sanitary T.

Reducing No-hub fitting makes it possible to go from 2-inch cast iron to 1½-inch plastic drain pipe.

example, if the horizontal distance from a connection under a sink to a point where the vertical section of the drain must pass down through the floor is exactly 2′, your pipe length will be a little less. The elbow that starts it on its downward path adds some overall length. Your best bet: check the chart for the distance the different fittings add. Then check the actual fitting by attaching it to the pipe and measuring the distance it adds. (Not all lines of fittings are exactly the same.) Then cut your pipe accordingly. Fortunately, there's usually — but not always — leeway in plumbing measurements. If you measure short in one place the chances are you can make it up by measuring long in another.

Drain-pipe diameters for all types of fixtures are usually specified by the local plumbing code. If there's no code in your area you can use 1½″ or 2″ pipe for basin, sink, and tub drains. The smaller size cuts costs. The larger one speeds draining and minimizes stoppages. Toilets connect to the soil stack with a "closet bend," usually 4″ diameter lead pipe.

The soil stack into which your new drain line will empty is most likely to be 4″ cast-iron pipe. In many new houses, however, it may be 3″ copper tube, if the local code permits. If it's copper, your connecting-in isn't difficult. When you have led your drain pipe to the point of connection to the soil stack (to ease final corrections leave the preceding drain pipe connection "dry" — held by friction tape), you're ready for the final step. The fitting, as shown in the drawings, will probably be a sanitary T or a Y branch. These fittings have rounded curves where they connect to the incoming pipe. This prevents solid matter from snagging or packing, and assures a fast, smooth flow.

If the stack is copper, simply cut out a section of the stack somewhat longer than the connection fitting. Then connect a length of stack-size tube to each end of the fitting to bring the total length equal to that of the cut-out section, allowing for the slip-on distances of the pipe and fitting. Now, when the fitting and the short sections of stack tube are soldered together the combination fits exactly into the space previously occupied by the cut-out section.

To link it all together, use slip couplings in the same way as described for connecting into water pipe along a straight run. The slip coupling can be pushed all the way onto the cut ends of the stack. Then the connection fitting assembly is set in place. The upper slip fitting is lowered over the connection at the upper end of the fitting assembly until its midpoint is at the point where the ends of the stack tube meet. (This after cleaning and fluxing the parts.) Then the slip coupling is soldered in the usual manner, as described in Chapter 5, Repairing Leaks. The new drain pipe may be joined to the fitting before the slip-joint soldering, but it must be cut precisely so that the replacement assembly (fitting and short tube lengths) will align exactly when set in the stack. To keep previously soldered connections in the fitting from softening when you apply heat for another one, wrap wet rags around the tubes close to the connections you want to keep from softening.

Before making slip coupling connections in a soil stack, however, be sure to check your local code. Some codes rule out slip couplings for this purpose, evidently with the thought that the coupling might not be properly centered with its midpoint at the juncture of the pipe ends.

If the soil stack is cast-iron pipe of the usual bell and spigot type with leaded connections, the job is more difficult. You are likely to face the job of disconnecting fixtures above the location of the connection fitting to be added. In this event the upper portion of the stack is then jacked up to permit a length of pipe to be removed. The lead joints can be drilled or chiseled free. This is then replaced with a shorter piece that, with the fitting added, makes up the same total length.

Use a No-hub sanitary drain fitting if the local plumbing code permits it. The No-hub fitting is far easier to install. This is the general procedure: The upper sections of the drain stack are supported to prevent them from dropping. The cuts in the stack are spaced about 1 inch farther apart than the overall length of the new fitting. Use a hacksaw to cut the cast iron. The new fitting is then moved into place and secured with No-hub fittings. It isn't difficult if you have sufficient space to work in.

To connect to a plastic drain pipe, use the same technique suggested for tapping into a plastic water pipe. You can either use a special fitting called a saddle, or you can slit a plastic sanitary T lengthwise using a hacksaw. Make a hole in the side of the pipe, as previously mentioned, but be extra careful to make certain the hole is slightly larger than the opening in the T and that there are no burrs so there is no possibility of muck hanging up at the hole and eventually blocking the drain.

Since you will have considerable plastic surface making contact and since there is very little pressure on the liquid in the drain system, you don't need clamps, but to be certain and to support the weight of the added-on drain pipe, it is advisable to use clamps.

In some cases it is possible to connect in at a clean-out fitting. To do this, remove the clean-out ferrule (the part that has the screw-in plug) and replace it with a sanitary T-Y fitting or Y branch. The clean-out ferrule must then be replaced in one connection of the new assembly so as to be directly in line with its former position. The new drain line enters the remaining connection. If your new drain line is to be connected to the stack in an expansion attic (as in adding a room and lavatory there) the job is made simpler by the fact that there are no fixtures above the connection point. So a connection between two lengths of the stack pipe can be separated, and the upper section of the stack lifted to permit the insertion of the necessary fitting. This, of course, entails loosening the flashing around the vent end of the stack above the roof.

There may be a problem, too, in the locations of the connections between lengths of stack pipe. If the connections are located part way up from the attic floor or the floor below you're likely to have plenty of work ahead. Your best bet: read the sections of your local code that pertain to the job you want to do. If it doesn't answer all your questions, get your remaining answers from the building inspector before you start any work. Then decide whether to tackle the project yourself or call a plumber. There's a good chance you can hire a plumber to do only the part of the job you don't want to undertake. You might, for instance, have the drainage work done professionally, and do the water-supply plumbing yourself.

A *complete new stack* is sometimes the best answer. This can be located in a wall of the new lavatory, for example. It can be led up through the roof at the top and vented just like the existing stack. And it can be led out through the basement wall just like the existing system. Outside, it can be connected underground to the existing house sewer line — if the type of pipe used for that makes connecting easier.

The extra stack is often the best answer (or the only permissible answer) when the new fixtures are to be located some distance from the existing stack. It will have to be of the same diameter as the existing main stack (soil stack) if a toilet is to drain into it. Regulations often allow a smaller diameter stack, however, if the only drainage involved is from sinks or appliances like automatic washers.

HOW TO CONNECT PLUMBING FIXTURES. The diagrams show the routes the water-supply pipes and the drains and vents must take to service the plumbing fixtures. At the fixtures, of course, these pipes must be connected to faucets, supply valves, and drains. Fortunately, this is often the easiest part of the job. As shown in the drawings, both drain and supply pipes are connected to the fixtures by methods and fittings not used anywhere else in the plumbing system. The reason: the design of some fixtures makes the usual connecting methods impossible, and others, that may require replacement parts at times, must be easy to disconnect.

Connecting a toilet to its drain pipe. As the toilet bowl in conventional types is bolted tightly to the floor, it is impossible to reach under it to tighten the final connection. And, as this connection may need to be separated in the future (as in making a repair or replacing the unit with a newer model) it would not be practical to require opening of the ceiling below to reach the connection. So the bowl seals on to the "closet bend" that connects it to the stack by a unique means.

The closet bend, the final section leading to the toilet bowl from the soil pipe, is caulked into the sanitary T fitting in the soil pipe by the same method used on all other soil-pipe full-sized connections — assuming you're using the usual cast-iron pipe. A brass ferrule in the soil-pipe end of the usual lead bend keeps it from collapsing during the caulking. The method used at the other end (where the bend comes up through the floor) varies, but the basic idea is the same. The floor end of the bend is connected to a brass flange that rests on the floor. (The end of the bend is trimmed so the flange seats snugly according to the method used.) In some cases the flange is connected to the bend in the same manner that the bend is connected to the sanitary T in the stack — by caulking with oakum and lead (Chapter 3). Or the flange may be of a type that permits the soft metal of the bend to be peened (hammered) over a beveled inner rim of the brass, then soldered. The type you use will depend on the local code.

Once the flange is secured to the bend it is snugged down against the floor. Some types are made in two parts joined by a threaded connection so that the upper portion of the flange can be simply turned to tighten it down. Others are one piece with screw holes for fastening to the floor and screw slots for the screws that hold the bowl down on the flange. The slots permit shifting the bowl position by rotating it to align with the wall.

Whatever the type of flange, the final step is the same. A sealing ring is laid on top of the flange to waterproof the connection to the bowl. Putty is sometimes used for this ring, but a ready-made wax ring (available from plumbing suppliers) does a far better and more permanent job, and it makes the work easier. With the wax seal in place, and the hold-down bolts extending upward from the flange to match the holes in the base of the bowl, you merely lower the bowl straight down. Then tighten the hold-down bolts to draw it tight on the flange and the floor. As you do this you squash the wax firmly into the circular recesses in the bowl base and in the flange, sealing the joint between the two.

Mounting the flush tank on the bowl is a relatively quick job. The drain outlet of the bowl, of course, should be such as to locate the bowl so the flush tank that rests on it can also be fastened to the wall behind it. (Tank mounting screws must go into wood framing members inside the wall.)

If the toilet is a new one, be sure you get a template (like a full-sized drawing), or a dimensioned diagram that shows you just how far from the wall the drain opening should be. New toilets may not require wall bolts.

If the toilet is a used one (as from a house-wrecker's yard) assemble it "dry" with the tank in place, propped against the wall, and trace the outline of the base of the bowl on the floor. Then you can make a full-sized template of the base (underside) and drain location. Newspaper is good for this. If you have any doubts about getting it accurate, allow for the toilet being located a trifle farther than necessary from the wall. It's then an easy matter to mount wood spacers behind the tank. You can run the tank-mounting bolts through them into the wall framing. If the whole unit is mounted too close to the wall, however, you may have a big correction job to do.

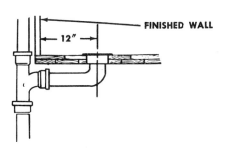

Important dimension in adding new toilet is shown here. Be sure toilet flange is located at correct distance from wall for toilet you buy, allowing for both bowl and tank.

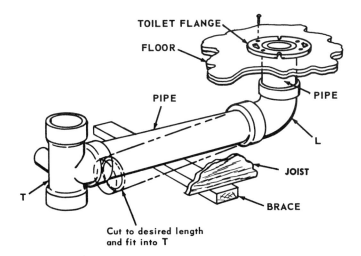

Cut to desired length and fit into T

To make sure all's well, assemble pipe sections "dry" with toilet flange, and set toilet in place for final check before soldering or lead-sealing connections. Temporary braces across joists can hold parts during try-fit.

Most types of flanges allow for moderate shifting of bowl around connection point for easy aligning with walls of room. (Slots in flange permit bolt movement until tightened.) After sealing gasket is in place under bowl base, however, try to position bowl correctly so it can be tightened straight down. Excess rotating movements can spoil the seal.

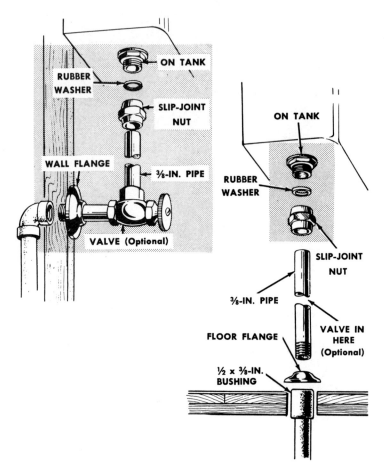

Supply pipe to toilet tank is connected as at left if pipe leading to it is inside wall. Connect as at right if pipe comes up through floor. Slip-joint connection is made last. Trim supply pipe carefully to fit from last threaded connection to slip joint.

Some flush tanks and bowls are designed for close-coupled mounting. The tank rests directly on top of the rear of the bowl with a conical rubber gasket sealing the water passage between them. Snug the two parts together with bolts provided for the purpose.

In other designs the flush tank is located several inches above the bowl and connected to it by a large-diameter chrome-plated pipe called a "spud." Conical rubber gaskets are tightened into both tank and bowl by large "spud nuts." To avoid chipping the porcelain, use a spud wrench for this. You can buy an inexpensive one from your plumbing-supply dealer.

The screws used for mounting the tank on the wall and for tightening the bowl down on the floor, in some flange types, are similar. One end is threaded like a wood screw, the other like a bolt. To screw these into wood you put *two* nuts on the bolt-threaded end and tighten them together. This locks them so you can drive the wood-screw end into the wall or floor by turning the upper nut with a small wrench. When the wood-screw threads are all the way in (smooth shank between them and the bolt threads), you can remove the locked nuts by simply screwing them apart. You need two small wrenches both for the locking and unlocking. Plumbing suppliers and hardware stores stock these screws and the washers that go with them.

If the toilet is to be mounted on a new cement floor, it's best to have all parts in place when the floor is laid, so that fastenings can be embedded in the cement. Otherwise, you can use a variety of masonry bolts for the purpose. A masonry bit in your power drill can make the necessary holes. The bolt expands in the hole for a tight grip. Hardware stores stock them.

The water-supply pipe on older flush tanks is usually connected to the bottom of the float valve by means of a slip fitting. The supply pipe leads to within a fraction of an inch of the threaded fitting on the bottom of the tank. A slip-joint nut containing a rubber washer slides over the supply pipe (something like the slip coupling mentioned for copper pipe earlier). When this nut is pushed up the pipe and tightened onto the threaded fitting in the bottom of the tank, it compresses the rubber washer against the pipe and the fitting, sealing the connection. That completes the job.

Modern tanks and the modern float valves that come with the tanks and can be used as replacement valves in older tanks are designed to accept Speedee fittings. Speedee fittings are illustrated and discussed in Chapter 5. If your present arrangement includes a slip-joint fitting that connects a 3/8-inch threaded pipe, you can install a special fitting that will couple the 3/8-inch pipe to the end of a 1/4-inch Speedee fitting tube and so install a new, modern float valve in an old tank.

Close-up of a Speedee fitting used to supply water to the bottom of a faucet. Curved portion of Speedee enters bottom of faucet shank. Flange nut holds it in place. Joint can be assembled and disassembled as many times as necessary without leaking.

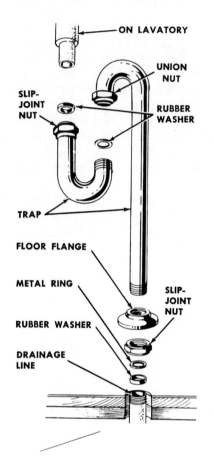

ON LAVATORY

UNION NUT

SLIP-JOINT NUT

RUBBER WASHER

TRAP

FLOOR FLANGE

METAL RING

SLIP-JOINT NUT

RUBBER WASHER

DRAINAGE LINE

Where drain pipe passes through floor under lavatory, it is connected like this. The union nut (near top of drawing) screws onto trap. A slight flare-out at the end of the drain pipe keeps nut from slipping off and acts as gripping shoulder for nut to turn on when tightening connection together. Rubber washer is squeezed between parts to form seal as union nut is tightened. Slip joint also uses rubber washer for seal, but does not have flared-out end on pipe, so it permits parts to be adjusted with some leeway before tightening.

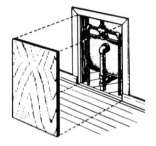

Bathtub trap is not accessible like lavatory trap, as it is under tub. If at all possible, provide removable wall panel in adjacent room to afford access to tub trap.

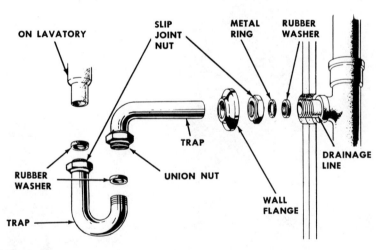

ON LAVATORY

SLIP JOINT NUT

METAL RING

RUBBER WASHER

TRAP

DRAINAGE LINE

RUBBER WASHER

UNION NUT

WALL FLANGE

TRAP

Where drain pipe is inside wall, connections are made like this. Connecting fittings are basically the same as in preceding drawing.

Wash basin drains (with trap) are connected to the drain pipe proper with a similar slip-joint arrangement, as shown in the drawing. The trap unit, however, is usually connected into the assembly with a positive flare joint arrangement that doesn't permit slipping up and down. If you take a careful look at the drawing you'll notice that the trap can be removed by screwing off both nuts, and simply pulling down on the U portion. You'll find some variation from this connection system, but the basic assembly and disassembly procedures are almost the same.

Sink drains are practically the same as lavatory drains, but larger. The same applies to washtubs in many cases, though the fittings may be galvanized rather than chromed.

Bathtub drains use the same connection system but lead to a different type of trap. The plumbing end of the bathtub should be against an inside wall so you can reach it from the other side to make connections, and when necessary, to clean the trap. A removable panel in the wall of the adjacent room makes the future servicing easy, even though the tub may have been installed before the walls were covered. If you omit the removable panel you may eventually have to cut an opening in the wall anyway if a stubborn stoppage occurs in the tub trap. If the tub is on the first floor, however, a basement ceiling panel may be the answer.

Faucet connections depend on the type and purpose of the faucet. Utility types like outdoor sill faucets are threaded or soldered to the supply pipe just like a pipe fitting. Lavatory and sink faucets, however, are usually connected to a final section of special supply pipe with a metal shoulder (larger than the pipe diameter) at the faucet connection end. A sliding nut on this pipe threads onto the faucet, drawing the shoulder tight against the faucet inlet and sealing the connection. Although there is some variation in design, the method of assembling these connections is always apparent.

In cases where water-supply plumbing faces unusual problems because of structural factors in the house, such as decorative walls you don't want to cut into, you can sometimes find your answers in combinations of pipe, tube, and fitting types—if your local code permits. And you'll find you can buy fittings to adapt practically any type of water pipe to any other, and to faucets and fixture inlets. For example, you can buy adapting connections that link small-diameter soft copper tubing to galvanized pipe, and you can buy similar adaptors to link the tubing to faucets. The interconnection of different metals in one system, however, isn't advisable where the water is on the acid side.

Modern faucets are usually connected to their water-supply pipes by means of a water-supply Speedee fitting. Like the Speedee used for connections to float valves, the water-supply Speedee is a length of $1/4$-inch copper tubing, usually plated. One end is expanded and fits into the end of the faucet shank. A flange nut that screws up on the shank holds the Speedee in place. No washer is used. Most often the other end of the Speedee is connected by means of a metal-ring compression joint, usually integral with a shut-off valve. But the tube end of the Speedee can be soldered or joined by means of a flare fitting just like any other length of copper tubing.

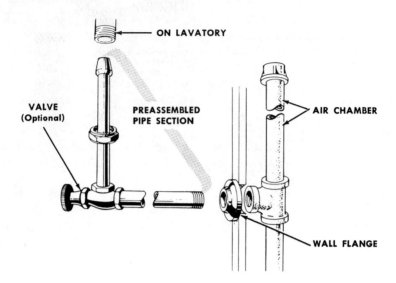

ON LAVATORY

VALVE
(Optional)

PREASSEMBLED
PIPE SECTION

AIR CHAMBER

WALL FLANGE

Supply pipes for lavatory faucets are connected like this. Conical end of supply pipe is drawn tight against faucet connection by tightening nut. Make this connection last. To prevent noisy "water hammer" when faucets are shut off suddenly, include vertical air chamber at least a foot high, as shown.

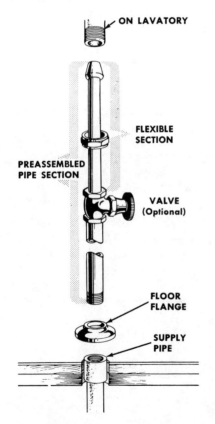

ON LAVATORY

FLEXIBLE
SECTION

PREASSEMBLED
PIPE SECTION

VALVE
(Optional)

FLOOR
FLANGE

SUPPLY
PIPE

Where water-supply pipe comes up through floor to lavatory faucets, connections are made like this. T and elbow may be used to provide air chamber (which must be vertical) behind water-supply pipe, as close to faucets as possible. Air chambers are most important where faucets are at end of long runs of pipe.

To connect a dishwasher or washing machine to a sink's drain pipe, the existing drain pipe is removed and replaced with a washing-machine T, as shown. Appliance drain pipe is then connected to the side of the T.

CONNECTING APPLIANCES. Clothes washers and dishwashers are considered appliances. The methods of connecting them are similar to those already discussed. A dishwasher requires hot water and a drain. A clothes washer needs both a supply of hot and cold water as well as a drain.

The easy way to connect either appliance is to tap into an existing water line (or lines) with a T. Then use an adapter to go from the T to a water-hose faucet, which is a plain faucet having coarse water-hose thread on its spout. Clothes-washer hoses are made to screw directly onto coarse-threaded faucet spouts. Dishwasher hose generally requires an adaptor between the hose end and the faucet. In some instances you will find it advantageous to use a shut-off valve rather than a faucet when connecting dishwashers.

Drainage for appliances is easily provided when the drain pipe can be hooked over a nearby sink. Another method is to extend the drain pipe by means of a water hose and lead the waste to a sink on a lower floor or to a floor drain.

If neither method is practical you can connect your appliance to a washing machine T, which is a special T that replaces the thin-wall brass tubing running from the sink to its trap. The T accepts the appliance's drain hose. The arrangement satisfies most plumbing codes, but isn't the best solution with a clothes washer, because washing machine waste contains a lot of clothes lint which combines with kitchen grease to quickly clog the drain. However, if you take precautionary measures (add lye or Drano every once in a while), clogging can be prevented.

7 | Outdoor Plumbing

IN AREAS where winters are cold, any outdoor water-pipe system, such as for lawn sprinkling, must be designed to permit complete drainage of all portions above the frost line. If the house has a basement this can usually be accomplished by pitching the main pipe downward toward its point of entry through the basement wall. As described in Chapter 6 (Adding a Garage Faucet), it is then possible to drain the system through a stop and waste valve inside. If the house has no basement, the main pipe may be pitched downward away from the house to a small pit about a foot or two deeper than the end of the pipe. With an ordinary sill cock on the pit end of the pipe and a stop and waste valve inside the house, the pipe can be drained into the pit. The pit need not be much more than a shovel-width square, as pipe of the size commonly used in these systems (about ³⁄₄″) doesn't hold a large amount of water in runs of usual length.

PLASTIC POOLS. The popular above-ground plastic pools require no underground or permanent plumbing, as they are usually filled with the garden hose. When they are to be emptied in the fall, however, the disposal of the 8,000 to 10,000 gallons the large ones contain may present a problem. Many communities do not permit emptying a pool into a storm sewer, and you can't drain it into a septic tank system or let it flood your neighbor's lawn.

The best bet for getting rid of the water is a small, electrically driven centrifugal pump. These are available from mail-order houses at relatively low prices compared to most swimming-pool accessories. Just run a hose from the inlet end of the pump to the bottom of the pool. Connect your garden hose or sprinklers to the outlet end, and switch on the pump. As the pool contents are spread over the lawn area there's no flooding and no runoff. And your lawn (also your evergreens) benefit from the end-of-season wet-down.

Do not add pool-conditioning chemicals to the pool water for several days before the lawn watering as some of them are harmful to plant life. However, these tend to dissipate over a period of several days. To be sure the water is safe for lawn use, try a sprinkling can full of the water on a small patch of lawn and wait overnight. Then, if all's well, start the pump.

GAZING POOLS for lawn decoration may have to be pump-drained if they are located on a level lawn. Where there is sufficient grade, an underground drain line from the low point of the pool to lower ground is better, as it may be left open during the winter to prevent accumulation of water from melted snow. (This is more important in pools with vertical sides that may crack from re-freezing, than in shallow bowl-shaped pools.)

A metal pipe coupling connected to a nipple and an elbow, as in the dia-

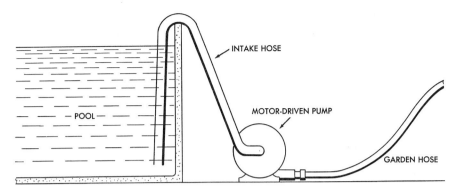

To drain above-ground swimming pool wtihout inundating surrounding lawns (when code forbids draining into storm sewer), use mail-order centrifugal pump, connected like this, and use pool contents to water lawn through sprinklers. To start it, fill intake hose with water, using pitcher and funnel. Hold thumb under open end until you have it well under water in pool, then start pump. Without this "priming" step, the pump may not pull water from the pool. It may take several days of sprinkling to drain pool, depending on size.

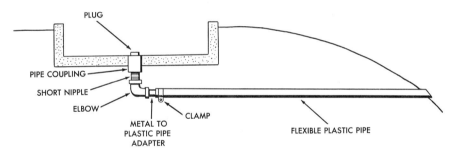

If you have shallow gazing pool on sloping ground, it's easy to rig a permanent drain pipe. Use 1 1/2" flexible polyethylene pipe. Adapters are available to connect it to metal pipe. All fittings shown are standard. When embedding pipe coupling in cement, tip it slightly to give entire drain line slight downward pitch. Plastic pipe can be flexed to increase pitch, if desired; can also be flexed to run around obstacles like boulders. It should emerge from ground where it won't be blocked.

gram, may be embedded in the concrete pool bottom. Flexible plastic pipe may be used as the drain line from that point on. (The same type used in deep well-pump systems.) In areas of extreme winter cold, rigid fiber pipe of 4" diameter may be advisable. The larger pipe diameter is less prone to blockage from a build-up of ice when freezing rain trickles through the pipe. This frequently occurs in winter when the drain line is left open to keep the pool drained. A standard pipe plug screwed into the embedded coupling in the pool bottom closes it in summer. The plug should be well coated with pipe compound or wrapped with plumber's teflon tape.

FOUNTAINS need be filled only occasionally in normal weather, as they re-use the same water. You can buy ready-made fountain-pump units at larger plumbing and garden-supply houses, or you can make your own with a motorized mail-order centrifugal pump. If you have a mechanical turn of mind you can do the job with an old washing-machine pump. The pump and motor unit should be protected from the weather by a waterproof housing that provides enough air circulation to prevent overheating the motor.

The most trouble-free systems have the pump located below the water level (though it can be a moderate distance from the pool), as this keeps the pump primed and ready to go. Avoid large pumps for this purpose, and adjust the pumping rate by changing the pulley sizes on motor and pump. If your fountain starts out as a geyser put a smaller pulley wheel on the motor, a larger one on the pump. If the fountain merely dribbles reverse the pulley-wheel arrangement.

. The motor current may be carried by direct burial (UF) type waterproof electric cable buried in a trench at least 18″ deep. Or it may be provided by a flexible waterproof cord (like an extension cord) which can be removed when the unit is not in use. The same electrical supply can be used to operate any night-lighting that may play on the fountain. If a permanent underground cable is to be used check with your local electrical code on the type of underground cable or other wiring required. If a temporary cord does the job it must run to an outside (weatherproof) outlet. Extension cords are not permitted to run through a building wall.

A point to remember if you have well water: Be sure your outside plumbing arrangements, whether for sprinklers, fountains, or swimming pools, do not use water faster than your well (*not* the well pump) can supply it. If you draw the level of your well water down below the intake of your pump system, you will run your pump dry. Many pumps cannot re-prime themselves automatically. So your dry-running pump may be seriously damaged or wrecked. This is most likely to happen when you leave sprinklers or a pool-filling job unattended.

Fountains re-use water constantly. Water lines should pitch slightly downward toward pump, to provide automatic "prime" and to permit easy draining, as in cold weather, through T's. Plugs in down-pointed branch of T's close them when fountain is in use. Fountain pump and motor units are available through large builder's supply houses, also mail-order houses. If immersible unit is used, no piping is required. But pump unit must be removed in winter, and fountain basin provided with drain.

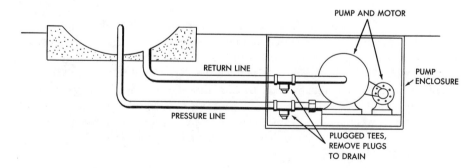

8 | Septic Tanks and Drainage Fields

YOUR GREATEST aid in installing and maintaining a septic-tank system is a general knowledge of how it works. Properly put together, the system should be completely trouble-free for a long time. Many have required no attention for ten years or more, and then only a professional pumping out.

HOW ABSORBENT IS THE GROUND? This is your first important question. The answer tells you how many square feet of drainage trench area you will need for each bedroom in the house. It is customary to figure this on the basis of bedrooms because this indicates the maximum number of people likely to live in the house. You can test the absorption of the ground by digging a foot-square hole as deep as your drainage trenches will be, then filling it 6" deep with water. Clock the time in minutes that it takes for the water to seep completely away. The chart tells you how many square feet per bedroom each seeping time requires. In any event, you must have at least 150 square feet. When you know the area required you can lay out the system.

Time required for water to fall 1 inch (in minutes).	Absorption area in sq. ft. per bedroom.
2 or less	85
3	100
4	115
5	125
10	165
15	190
30	250
60	330
Over 60	Special design using seepage pits.

HOW THE SYSTEM WORKS. In operation, a sealed-joint pipe (usually 4" diameter) carries all sewage and other plumbing drainage from your house to the septic tank. The inlet of the tank is slightly higher than the outlet. So, for each amount of fluid that enters, an equal amount flows out into the absorption field. To prevent the incoming sewage from flowing directly through the tank and out the overflow, both the inlet and outlet are fitted with down-pointing sanitary T's or shielded by baffle plates. This way, the incoming flow is directed downward to the bottom of the tank, while the outgoing flow is drained off from the top of the tank.

Bacterial action takes place at all levels in the tank, breaking down the solids to liquid, gas, and mineral sludge. The liquid is what flows into the drainage field. The gas drifts back up the sealed-joint pipe to the house, then escapes up the vent stack to the open air above the house. The sludge settles to the bottom of the tank — and this is what must occasionally be pumped out by firms operating trucks designed for the work. The sludge does not break down further to gas or liquid. But it accumulates very slowly compared to the amount of solid matter and liquid entering the tank.

Household chemicals like strong bleaches and detergents retard the necessary bacterial action by killing considerable amounts of bacteria. If the chemical inflow is not excessive, however, the tank continues to perform effectively. But large-capacity automatic washing machines and similar tank-overloading appliances can cause trouble. First, by sending large volumes of water into the tank each time they empty, they churn up the solids that are still being broken down by bacterial action, and they churn up the sludge. When these products flow into the drainage (absorption) field, they clog the pores of the soil and reduce its rate of absorption. Meanwhile, the household chemicals slow the bacterial action. So, in severe cases, the liquid effluent from the tank, not being absorbed by the earth, finds its way to the surface and overflows onto the lawn. As this can easily be prevented in most cases, the home-owner, not the tank, is to blame.

HOW TO AVOID TROUBLE. Whether you are installing a new septic-tank system or trying to improve an old one, there are a number of ways of dodging the kind of trouble just mentioned. (It's probably the most frequent septic-tank problem.)

You can switch to appliances that are designed to economize on water. To do this, find out from the manufacturer of your present appliances just how much water they dump into your drainage system each time they go through a full series of operating cycles. Then look for replacement equipment that reduces this volume of outflow in a *major* way.

If you are already using water-saving appliances but still have effluent overflow on the surface of your septic-tank absorption field, you may have enough property to separate your sewage (toilet) system from your other drainage. This is done by diverting the drainage from all fixtures (and appliances) *except* toilets to a separate main drain. This drain may be led to "dry wells," or drainage pits. These are usually walled with uncemented masonry blocks to prevent the sides from caving in. As only a very slight amount of solid matter is carried by the drainage from sinks, wash basins, and tubs or washing machines, the fluid in the pit has little tendency to clog the seepage spaces between the masonry blocks.

If there is a convenient and readily accessible place in the house for a "grease trap," it may be worthwhile to install one. Most of the grease in the drainage, of course, comes from the kitchen sink and dishwasher. Over a period of time it coats the walls of the dry well, clogs the soil, and slows absorption. But to be useful in preventing these troubles, the grease trap must be where it can be cleaned easily. In general, this consists of removing the cover, scooping out the accumulated grease, and disposing of it in the garbage can.

A giant-sized septic tank is another measure often used to minimize field-clogging problems. The greater tank volume, and usually greater distance from inlet to outlet plus greater depth, reduces the outflow of churned-up solids. But the household chemicals still have some effect on the bacterial action if large amounts of them enter the tank. And the excessive amount of drain water sent into the tank by some washing appliances can still saturate the absorption field. So, if you have the space, you may have to enlarge the field. Or take your wash to a laundry.

Just which corrective or preventive measures you will have to use may also depend on your local code. Some codes require that all drainage from all fixtures be piped to the septic tank. Others recommend that sewage from toilets and drainage from other fixtures be handled separately, as described earlier. So check fully with your local code and your building inspector before you plan a new system or modify an existing one.

LOCATION OF TANK AND FIELD. Requirements vary with the state and locality on the minimum distances between the various parts of septic-tank system and the house, well (if any), and property lines. If there's no local code to guide you, write or phone your state health department for a copy of the state regulations that apply. Or base your planning on the codes of nearby areas. Recommendations in a typical area call for a minimum distance of 5' between the house and the septic tank, 50' between tank and well, 100' between the absorption field and the well, and 10' between the perimeter of the absorption field and any property line. As many code areas call for greater minimums than these, however, your best bet is your state code if you have no local one. As to the distance from septic system to well, don't think of reducing it because the well is steeply up grade from the septic system. Underground strata along which the water flows beneath the surface may actually slope in the opposite direction.

INSTALLING THE TANK AND DISPOSAL FIELD. This job begins with the selection of the tank. Your local code may tell you exactly what size and type you are permitted to have. Otherwise the basic facts are your guide. You have a choice between welded steel and cast concrete tanks unless you build your own. If you build your own you should use solid (not hollow) masonry blocks, and "butter" the inside and outside of the job with a layer of cement to assure watertightness. In general, you'll save money or break even by buying a ready-made tank, and you'll be sure it's tight—at least initially. The metal tanks are lower priced and last for decades in noncorrosive soil. In other soils, however, their useful life may be considerably shortened. You can judge by the performance of metal tanks in the immediate area. If you want to cut costs and play it safe at the same time coat your metal tank with a heavy metal-protecting preparation and let it dry thoroughly before setting the tank in the ground. Asphalt-base roof coatings are often used for this purpose.

Digging the hole for the tank. The depth of this hole is particularly important. It should be such as to provide a downward pitch of the sealed pipe from the house to the tank of about $\frac{1}{4}''$ per foot of run. A flatter pitch is likely to cause clogging. Too steep a pitch can cause the water-borne sewage to surge into the

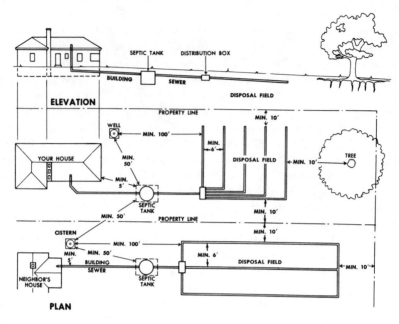

Typical minimum distance requirements applied to two common septic-tank and absorption-field systems. Note that drain from kitchen sink is separate line when grease trap is used. Shape disposal field to keep tiles at least 10′ from trees, as roots grow toward field's moisture and can clog tiles.

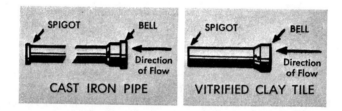

Direction of flow through any type of bell-and-spigot pipe is always as shown.

tank and churn the contents. Codes usually require cast-iron pipe at the start of the house-to-tank run.

The other factor involving the depth of the tank hole is the depth of the disposal field tiles — as this part of the system actually begins at the tank outlet. The outlet is usually 1″ to 3″ lower than the inlet, so the tank must be carefully leveled in the hole. Otherwise tilting endwise might cancel out the difference between inlet and outlet levels. The top of the tank should be at least 1′ below sod level, but not more than 3′, as it must be accessible for servicing — lid removal and pump-out. Some tanks have a small removable lid in the top, others have a completely removable top. (Tank-cleaning people need an opening big enough to stir the contents with a shovel so that both solids and liquid are drawn out completely by the cleaning pump.)

The disposal field. In a typical layout, sealed-joint pipe runs from the tank outlet to a "distribution box" from which lines of drainage tile radiate. The drainage tiles are loose-jointed or perforated to permit the liquid tank effluent to seep out into the ground. Your local code may or may not specify the distance from the tank to the distribution box. In any event, it's wise to locate it several feet from the tank so as not to concentrate effluent around the tank itself.

The tiles that release the effluent to the soil may be made of cement, kiln-fired clay, or impregnated fiber. Your local code may specify which you are permitted to use. The cement tiles are the lowest priced and are approved in many areas. They have established a good performance record over the years, too. The kiln-fired type are more widely approved, and also have proven long-lived. The fiber form (like Orangeburg pipe) is actually sealed-joint pipe with holes drilled through it at close spacing on opposite sides. As it comes in lengths up to 10' it simplifies the disposal-field job. But be sure you buy a quality brand, as some fiber pipes have deteriorated and cast unjustified suspicion on even the top grades. The better brands, however, have established an excellent record for both durability and strength.

Trench details. Your first trench leads from the tank to the distribution box. It should be just deep enough to take the sealed-joint pipe from the tank to the distribution box. The box can be bought ready-made at many masonry-supply outlets. Just specify the number of connecting openings you need.

The trenches for the loose-jointed or perforated tile or pipe must be slightly deeper so that about 6" of crushed stone can be laid in first to form an absorbent bed for the tile. The crushed stone is usually designated as 3/4" size (unless your code specifies otherwise), as that size doesn't clog readily. The tile is then laid on top. If you are using loose-jointed tile, however, grade boards should be embedded in the crushed stone. These boards are simply lengths of the cheapest lumber you can buy (usually 6" sheathing boards) to provide an even base to give the tiles their downward pitch. The boards are laid on edge in the crushed stone. You can establish the downward pitch (about 1/4" in 6') by taping a carpenter's level to a straight 6' board with a nail head protruding 1/4" from its lower edge at one end. Set this end on top of the grade board, pointing away from the distribution box, and work the grade board into the crushed stone until the bubble in the level is perfectly centered. Or you can use a taut string and a "line level."

Lay the tiles (usually 1' or 2' lengths) on top of the grade boards and mound up enough crushed stone on each side to hold them in place. The spacing between the tile ends should be from 1/4" to 1/2". Cover the top of this gap with tar paper or, better, lightweight roll roofing. This keeps earth from sifting into the tiles at the gaps where the effluent seeps out. After covering the gaps, cover the tiles with about 2" of the crushed stone (unless your local code specifies otherwise) and cover the top of the crushed stone with a layer of newspaper or a 2" layer of straw. This prevents the earth (which is shoveled in next) from sifting down into the crushed stone. By the time the paper or straw rots away the earth above it will have packed enough not to sift into the crushed stone.

The overall depth of the trench should seldom be more than 30" to 36" (be-

As inlet and outlet heights are only 1" different in many metal tanks, be sure tank is level in earth. If tilted so inlet and outlet are at same height, or if inlet is lower than outlet, liquid may back up in inlet pipe and prevent sewer gases from escaping through vent stack. With top off of tank, you can tell outlet side because metal baffle on that side extends deeper.

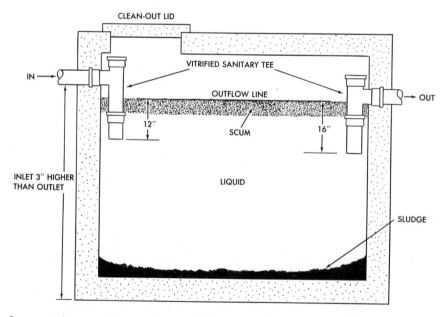

Concrete tanks are usually built with inlet 3" higher than outlet. Sanitary T's usually take the place of the baffles used in metal tanks. Either T or baffle must be used to prevent scum from flowing into pipes and possibly causing stoppage. T's or baffles extend below scum level into clear liquid. Scum and solids break down constantly into gas, liquid, and small amount of solid mineral sludge that settles to bottom. When this sludge builds up enough to be churned up into outlet pipe, tank should be pumped out. Local tank installers can tell you how much time to allow before pumping tank—commonly around two to four years, depending on tank size and amount of sewage it handles.

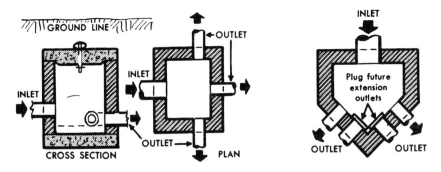

Distribution boxes can be bought ready-made, or you can make your own from concrete to match the number and direction of branches needed. Inlet is higher than outlets, as shown. If there's land enough to expand the absorption field (in case it should become necessary) it's wise to include plugged outlets in distribution box so new lines can be added easily.

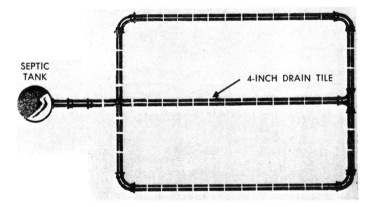

Loop disposal field like this uses a drainage pipe "cross" fitting (left) and T fitting (right) instead of distribution box. But check with your local code before using. It's best suited to fairly level ground.

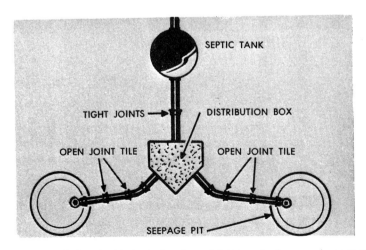

Seepage pits are used where land isn't suited to disposal field layout—for example, on steep slopes.

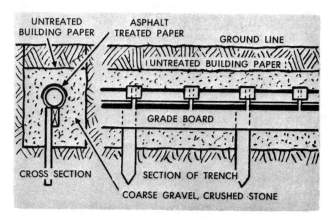

Disposal trench construction. Grade boards can be installed and set at correct pitch angle before crushed stone is put in, if stakes are used as supports for boards, as shown. Cover 1/4" to 1/2" gaps between tiles with asphalt paper or lightweight roll roofing.

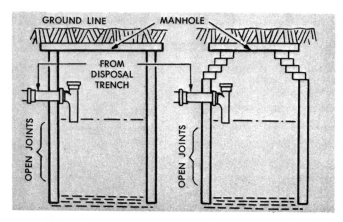

Walled seepage pits for use where disposal field isn't practical. Concrete block laid with uncemented joints is easiest wall material to use. If pit must be built near trees it may be made without wall—simply a hole in the ground filled with loose, crushed stone. Then roots can enter pit without damaging it. Roots take on a large amount of moisture, improve performance of pit.

fore laying crushed stone) as the root system of the lawn grass actually plays a considerable part in carrying off the moisture absorbed by the soil. The grass is literally watered from below. So don't lay your tiles so deep the grass roots can't reach the moisture. (The disposal field, however, can be put into operation before the lawn above it is established.)

The ends of the tile runs may be blocked closed with a flat masonry block or filled in for about 6" with crushed stone to prevent excessive end seepage. In general, no run should be longer than 100', and runs should be at least 3 times the trench width apart—but not less than 6'. And, of course, the field should be located where no cars or trucks will have to pass over it. It's a good idea, too, to plant some small perennial flowers at the location of the tank pump-out opening and at the location of the distribution box. This saves a lot of exploratory digging if repairs are ever necessary, and whenever the tank is to be pumped out. But don't mark the spots with a tree. Instead of plants, you can use lawn ornaments (like a bird bath) if you prefer.

Where there isn't room for a disposal field, or where the terrain contours make it impractical (as on steep slopes) the alternative is usually one or more dry wells, as described earlier. If these must be located near trees or wooded areas, however, it's usually wise not to use the type with uncemented masonry block walls, as roots may wedge in and dislocate the blocks. Instead, fill the dry well with crushed stone like the trenches, and with rocks up to orange size. The roots can work through this mixture without disturbing it, and they can play a sizable part in disposing of the water.

GENERAL TIPS. Before you buy a septic tank make sure it's a type approved for your area, and find out how it will be delivered. Concrete tanks are usually delivered by a special truck equipped with a crane to lower the tank into the hole. But you must have the hole ready for it. (The truck will usually wait long enough for you to check the level of the tank—and hoist it a little if you have to toss some earth under one end to level it.)

Metal tanks of average size may or may not arrive on a crane truck, as they are lighter in weight. It's not usually too much trouble for two people to roll one of these tanks to the hole location, but you'll be wise to build a wood tripod and use a block and tackle to lower it in the hole. Find out the tank weight in advance so you'll know what you have to handle. A typical 300-gallon model weighs about 270 pounds, a 750-gallon model about 500 pounds.

Whatever type of tank you use, it's a good idea to fill it at least half full of water as soon as it's in the ground. If you don't, a sudden increase in the ground water level (as from heavy rain) can actually float the tank upward—even push it up through the earth above it, after you've covered it.

If, shortly after you put your tank and its disposal system into service, you encounter a major stoppage that causes back-up into fixtures or overflowing toilets, the trouble is likely to be easily cured. It's most likely to occur where glazed tile with cemented joints has been used underground. A relatively small snag of hardened cement inside the pipe can catch solid matter and cause a blockage to build up quickly. A plumber's tape or a long rigid rod let's you "feel" the snag and click it off after the pipe is emptied. Your best bet: avoid this nuisance by pushing a wad of cloth through the cemented pipe run while the joint cement is still soft. Use a stock lumberyard 1-by-2 to push the wad—and don't make the wad a tight fit. If you push it from the house toward the tank (you can do it a few sections at a time, as you lay the pipe) it will smooth any oozed cement in the direction of flow anyway. It's a good practice to follow the same procedure in the pipe leading from the tank to the distribution box.

WHERE A SEPTIC TANK CAN'T BE USED. In areas where the soil is wet or nonabsorbent, or where a shallow ledge of rock creates a disposal field problem, the conventional septic tank system may not be feasible. In remote locations, too, where vacation cottages are likely to be situated, it may not be practical to bring in the materials necessary for a septic tank and field. The best substitute under these circumstances is likely to be a completely different type of sewage-disposal device called the "Destroilet," developed in 1960 and now widely used in problem situations. This unit requires no actual plumbing. Essentially, it is an incinerator type of sewage-disposal unit combined with a toilet. Instead of a water-supply pipe and a drainage line, it requires a gas-feed line and a flue pipe like a furnace. When the toilet lid is closed a 2100-degree gas flame converts all sewage matter to carbon dioxide, water vapor, and inert ash. No chemicals are required, and in a family of six to eight people, ash-removal should usually be necessary only about once a month. It disposes of all toilet waste including disposable diapers and feminine hygiene items. It is approved by the American Gas Association for use in both residences and mobile homes. The manufacturer: La Mere Industries, Inc., Walworth, Wisconsin.

9 | How Heating Systems Work

THE TYPE of heating system in a house often depends on the age of the house. If you buy a really old home it may be heated by a gravity hot-air system or a gravity hot-water system. Or, in a few instances, it may have steam heat. Newer houses are likely to be kept warm by forced warm air, hydronic (forced hot water), or electric heating systems, plus a number of modern variations. All of these systems, new and old, can do a commendable job if properly installed. And the newer types, as might be expected, usually provide more comfort and economy. But the difference between the old and the new isn't always as great as you may imagine. So, before planning a change, it pays to weigh all the facts.

THE GRAVITY HOT-AIR SYSTEM is one of the oldest and simplest. The furnace, essentially, is just a small metal box inside of a big one, with a sealed flue pipe leading from the inner box out through the bigger one to the chimney. Build a fire in the little box and you heat the air in the space between it and the bigger box that surrounds it. The heated air rises to the top of the big box and enters a tip-top chamber called a "plenum" to which hot-air ducts are connected. These big, up-sloping pipes carry the rising hot air to the rooms in the house above. The hot air floats up through the registers in the rooms, heats the rooms, and settles to the floor as it cools and becomes heavier. Once at floor level, it flows down through another set of ducts that carry it back into the bottom of the furnace. There, it picks up new heat and rises to the rooms again. The cycle continues as long as there's a fire in the furnace — and the fire may be fueled by coal, oil or gas.

There are no moving parts in the circulating system so it works in silence, but it has some drawbacks. Because the difference in the weight per cubic foot of hot and cool air is only a tiny fraction of an ounce, the ducted air doesn't move with much force, so it can't push its way through a conventional dust filter without slowing down too much for efficient heating. Hence, gravity hot-air systems are usually unfiltered.

The gentle flow also makes it necessary to keep the hot-air ducts short to avoid too great a heat loss along the way. So the hot-air registers are usually at the *inside* walls of the rooms — closest to the furnace situated in the center of the basement. And the return air registers are at the *outside* walls. (They have to be on the opposite side of the room in order to draw the hot-air flow across.) Thus, the hot air floats past the warmer inner walls first. Then, as it cools, it descends past the colder outside walls and windows and enters the return registers for its downward glide to the furnace again. As a result, you may find the outside areas of some rooms a little chilly on very cold nights. Also, the easy, natural air flow may make the unit a bit slow to respond to the thermostat.

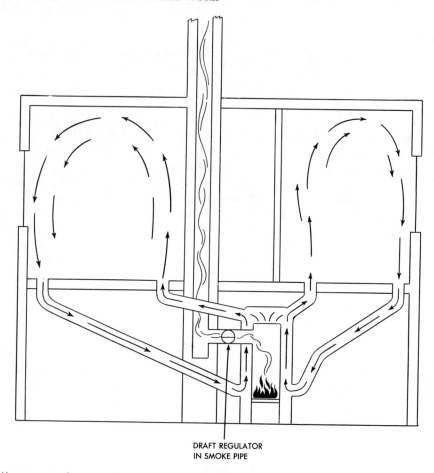

DRAFT REGULATOR
IN SMOKE PIPE

How a gravity hot-air system works: Hot air, being lighter, rises to the rooms, cools as it heats the rooms, then slides down the return ducts to the furnace for reheating. Circulating force is very small, so distant corners of rooms may sometimes be a little chilly. Heating ducts, which must be kept short, are usually at inner walls, closest to furnace.

In a well-built old house the sum total of these effects may not be noticeable enough to make a complete heating system change worth the expense. Your best guide is your own comfort.

FORCED WARM AIR, one of the most popular modern systems, is very similar to the old gravity type. But the return (cool) air ducts are connected together before they get back to the furnace — and a blower powered by an electric motor sucks the cool air down the ducts and shoots it into the base of the furnace for a high-speed trip past the hot fire box and back to the rooms. (We now call the box with the fire in it a "heat exchanger.")

The high-velocity air stream carries the heat away from the hot furnace surfaces and into the rooms much more rapidly — just as fan-driven air whisks the heat away from an air-cooled engine, as in the family Volkswagon. So you

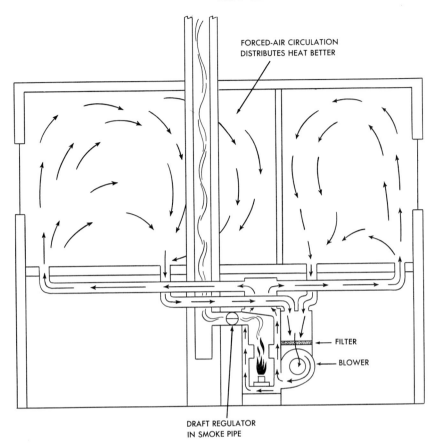

FORCED-AIR CIRCULATION
DISTRIBUTES HEAT BETTER

FILTER

BLOWER

DRAFT REGULATOR
IN SMOKE PIPE

How a forced warm-air system works: Motor-driven blower drives air through ducts to rooms, so heat-supply ducts need not be short. Hence, warm air can enter room at outside walls, which is best for efficient heating as outside area of room is likely to be coldest. Return air is drawn back by blower, then driven into furnace for reheating.

can get much more heat from the same size furnace, or just as much heat from a smaller furnace. And, as the forced air stream can travel in any direction, you can take the furnace out of the cellar and put it anywhere you please — including the attic. The power-driven blower will shoot the heating air down, up, sideways, anywhere you want it. Basically, that's the forced-air story, although there are variations in detail.

Both the gravity hot-air and the forced warm-air system can be improved by raising the moisture content of the air that is circulated. Moist air carries more heat than dry air at the same temperature. Also, the addition of moisture to the hot-air stream prevents damage to furniture and relieves dryness in one's nose and throat.

One way to add moisture to a hot-air heating system is to install an automatic

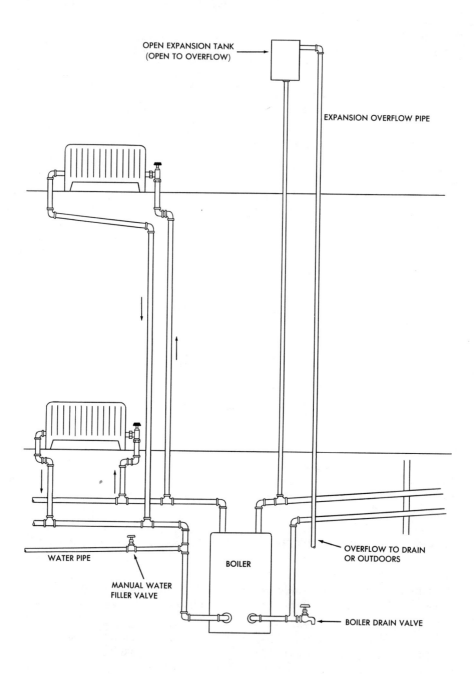

Layout of a typical old gravity hot-water system. Water rises and falls as shown by arrows. If water expands beyond capacity of system, it simply flows out of attic expansion tank and down to cellar drain, sometimes even to rain gutter of attic roof.

humidifier. This is accomplished by cutting a hole in the side of the plenum or the major hot-air duct and inserting the humidifier.

The humidifier is held in place by means of self-tapping metal screws driven into the side of the plenum or duct. The water-feed tube is connected to the nearest cold-water pipe.

GRAVITY HOT-WATER SYSTEMS are similar in principle to their gravity hot-air relatives. A fire under the boiler heats the water. The hot water, like hot air, rises to the top of the boiler and flows up into a supply main (large pipe) that slopes upward gradually around the basement perimeter. Smaller pipes extend upward vertically from the supply main to one end of each radiator, to let the hot water flow up into them. At the opposite end of each radiator, another pipe (connected at a low point) carries the cooled water down to the "return main"—a large pipe that slopes downward around the basement perimeter and eventually leads into the bottom of the boiler. Thus, water heated in the boiler floats up into the radiators and heats the rooms. And, as it heats the rooms, it cools off and becomes heavier—just like the air in a gravity hot-air system, flowing down the return pipes and back into the boiler to start all over again.

As the water heats and cools more slowly the system provides a somewhat more even-temperatured heating than the hot-air counterpart. But, unlike the hot-air system, it must be completely drained if the house is to be left unheated. And, of course, it must be refilled before the heating system can be fired up again.

Gravity systems can be improved in efficiency and response time by adding a circulator pump to the water circuit. In most instances it is a much wiser course than that of replacing the old system with a new one. The steps necessary are neither difficult nor complicated, but unfortunately there isn't sufficient space in this volume to describe them.

THE HYDRONIC SYSTEM is simply a forced hot-water system. The supply main is connected to the top of the boiler and the return main to the bottom, as usual. But the water is sent streaking through the system by a centrifugal pump like those that drive the water through an automobile radiator. So, like the forced hot-air units, the boiler can be smaller, the pipes can be smaller, and the boiler need not be in the basement. The method of feeding the water through the radiators, however, varies in several basic ways.

The series loop is the simplest of the radiator-supply arrangements. Hot water leaving the top of the boiler through the supply main shoots straight through each radiator in succession, through the intervening lengths of pipe, and on around the circuit until it returns to the bottom of the boiler. The radiators are actually sections of the supply main—just as if they were lengths of the pipe itself, but fitted with radiating fins. So a minimum of fittings and material are required. But, since the radiators are integral parts of the supply main, no radiator can be shut off. As each radiator is a section of the main circulating system, shutting off any one of them would stop the flow through the entire system. So all radiators heat up and cool off together, as the central heating unit is turned on and off by the thermostat. Thus, the series loop is suited mainly to compact houses that do not require room-to-room heating adjustments.

The one-pipe system consists of a circuit of pipe leading out of the boiler top and returning to the bottom of the boiler, like the series loop. But the radiators are not an integral part of it. At each radiator location a branch pipe carries hot water from the circulating main to the radiator inlet. Another branch pipe carries the return water from the radiator outlet back to the circulating main. A shut-off valve at the inlet allows you to regulate the amount of water entering

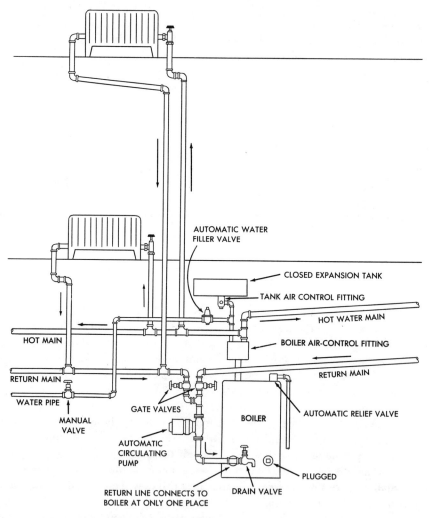

When gravity system is converted to forced flow (hydronic) you take away open expansion tank and substitute closed expansion tank that allows for expansion with cushion of air in top of tank. Automatic relief valve, also used with some gravity types, takes care of emergencies. All controls are made automatic. Return lines are joined together so that they enter boiler at only one place.

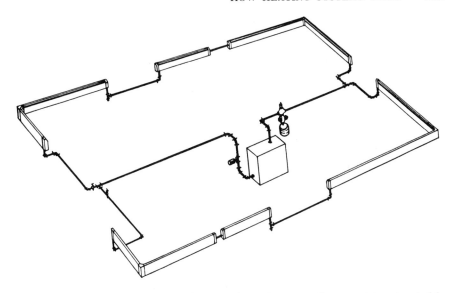

Series loop system has no heat-control valves on radiators because radiators are integral part of the mains. If one was shut off, flow would be stopped through entire loop of main.

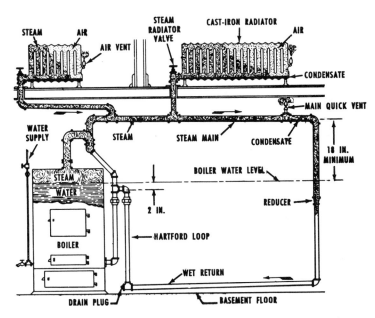

How one-pipe steam system works: Steam from boiler rises through main and branch pipes to radiators, pushing air ahead of it. Air escapes through automatic air vents on radiators, but vents shut as soon as steam hits them. As heat from steam radiates into room, some steam cools and condenses to water—which runs back through branch pipes and mains to boiler. In branch pipes the water running down actually passes steam coming up. In the main, both travel in the same direction. The Hartford Loop introduces steam pressure into the return line to equalize the pressure through the system, so the boiler can't lose its water by forcing it out into the return line.

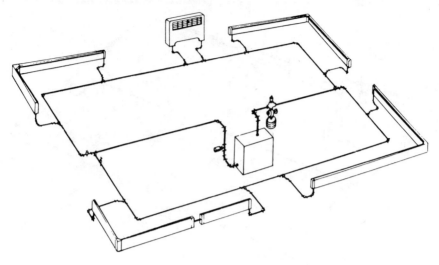

One-pipe system has branch pipes running from main to both ends of each radiator, with heat-control valve at inlet end of each radiator. If radiator is shut off, water continues to flow past it through main. Forced-flow T detours part of main flow through radiator when valve is open.

the radiator, or turn it off altogether. If you turn it off the water simply flows through the main instead of the radiator. A special T fitting (called a forced-flow T) is used where the return pipe from the radiator connects back into the main circulating pipe. This T has a little scoop facing away from the flow through the main, as shown in the diagram. This causes it to suck water through the radiator when the inlet valve is open — but it doesn't block the flow through the main when the valve is closed. The advantage of the one-pipe system, of course, is individual room heat control. But it requires more pipe, more fittings, and more installation work. As it combines central control (as with the series loop) with room adjustment, however, it is a very popular system.

The two-pipe system is really a gravity hot-water system with a pump driving the water through it. The hot-water supply travels to each radiator through a main and a branch pipe. And the cooled water leaving the radiator flows back to the boiler through a branch pipe and a return main — so it doesn't mix with the hot water in the supply main. Thus, the temperature difference between the first and last radiators along the supply line is considerably less than with a one-pipe system. Thus the two-pipe arrangement is a good one for large, rambling houses.

ZONE HEATING. Possible with either forced-air or hydronic systems, its purpose is to send heat only to the section (zone) of the house that requires it, and not to sections that are already warm enough. It is frequently required in split-level homes where the heat tends to drift upward through the large stair passages. Under these conditions the upper levels remain warm while the lower levels cool off. Supplying heat to both levels would make the upper level too hot. So the house is divided into zones, each with its own thermostat. Then, if

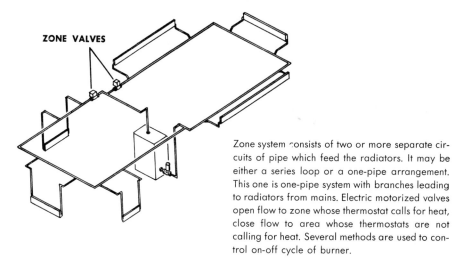

Zone system consists of two or more separate circuits of pipe which feed the radiators. It may be either a series loop or a one-pipe arrangement. This one is one-pipe system with branches leading to radiators from mains. Electric motorized valves open flow to zone whose thermostat calls for heat, close flow to area whose thermostats are not calling for heat. Several methods are used to control on-off cycle of burner.

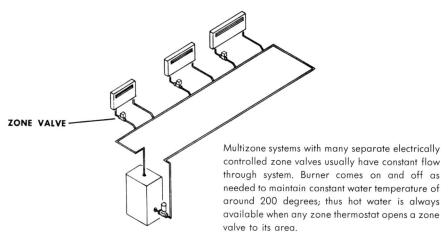

Multizone systems with many separate electrically controlled zone valves usually have constant flow through system. Burner comes on and off as needed to maintain constant water temperature of around 200 degrees; thus hot water is always available when any zone thermostat opens a zone valve to its area.

the lower level's thermostat signals for heat and the upper thermostat does not, heat is directed to the lower zone only, and blocked away from the upper zone. In a forced warm-air system this is accomplished by electrically operated shutters (like louvers) in the ducts. In a hydronic system (forced hot water) it is done by electrically operated valves in the supply pipes.

All zone systems are designed for complete flexibility. They can supply heat to any of the zones (often there are three or four) singly, to several at a time, or to all zones at once, according to the signals received from the various thermostats.

ELECTRIC HEATING. In its usual form, this system employs a number of individual room-heating units, commonly of the baseboard type. The heat comes from a resistance wire like those you see glowing in your pop-up toaster. But you don't see the glow because the wire is surrounded by high-temperature electrical insulating material, and enclosed in a metal tube like the rod-type burners on an electric range. In heating units, the tube is usually set in a larger finned shell. This, in turn, is covered by a metal baseboard housing. Each set of room units is controlled by its own thermostat. So independent room temperature control is possible if the room doors are kept closed. Although central temperature control is possible with this type of system, it is seldom included as it brings a considerable increase in initial cost.

Because of the ease with which electric cable can be led through the house structure, this type of electric heating is one of the simplest of all to install. (The cables are usually sized for 230 volts, the operating voltage of most baseboard electric-heating units.) However, as it costs considerably more to produce a given amount of heat with electricity than with oil or gas, extra-heavy house insulation is required to bring heating bills to a reasonable level. Electric Heating Industry recommendations call for 4″ of insulation in walls, 6″ in the ceiling, and 2″ in the floor. (As 2-by-4 wall studding allows space for only 3⅝″ of insulation, wall insulation rated as "equivalent" to 4″ can be used.) A vapor barrier, storm windows, and weatherstripping also should be used. In new home construction, however, masonry costs can be reduced as the electric heating system requires no chimney.

Central electric heating units are also available in both warm air (forced) and hydronic types, as are radiant heating systems using ready-made panels or heating cables embedded in walls, floors, or ceilings. In general, however, the popular baseboard system is the easiest to install and maintain.

HOW GAS AND OIL BURNERS WORK. Because of the difference in behavior between the two fuels, gas burners and oil burners differ greatly in their basic design and in their safety features. Either one, however, can be used with all conventional warm-air or hydronic systems.

The gas burner may have a single jet that shoots a flame against a deflecting surface to spread it over the surface to be heated. Or it may have a multi-jet arrangement like the burners on a gas range. In either type the main jet or burner is usually lighted by a pilot flame. And the pilot flame does another very important job: it heats a device called a "thermocouple" which converts the heat to a very small electric current. This tiny current, however, is sufficient to hold an electrical valve *open* in the gas supply line. If the pilot light goes out for any reason, the thermocouple cools off, the electric current stops, and the gas supply valve *closes*. Then the main gas burner can't be turned on (even though the wall thermostat calls for heat) until the pilot light is relighted. So, if the pilot light isn't there to light it, gas can't escape.

To prevent the pilot light from being blown out by chimney downdrafts (wind blowing down the chimney), gas heating units have a "draft diverter" either in the unit itself, or in the vent pipe leading to the chimney. (The smokepipe chimney system of a gas heating unit is commonly referred to as a "vent".) In one familiar design, a draft diverter is something like one funnel fitted inside of another, with air space between. Air flows easily from one spout into the other when moving from the wide portions of the funnels toward the narrow portions. But when the flow is reversed (as in a downdraft) most of the air escapes between the funnels. Thus, a chimney downdraft escapes into the furnace room instead of blowing into the furnace, thus protecting the pilot flame.

The oil burner in most oil-fueled central-heating units is a "gun" burner. (The name is generally attributed to the cannon-like form of the metal barrel through which the air and fuel oil enter the combustion chamber.) A single electric motor runs both the oil pump that feeds fuel oil to the spray nozzle at the firebox end of the gun, and the blower that drives the air stream in which the oil spray burns.

Instead of a pilot light, the oil burner uses a high-voltage electric spark to ignite the fuel. As soon as the burner starts, the spark streaks directly through the air-oil mixture being emitted from the firebox end of the burner barrel, and in many types, continues until the burner shuts off at the end of its heating cycle.

In the rare event that the fuel-air mixture fails to ignite, however, the burner shuts off automatically in a matter of seconds. So oil does not continue to spray into the furnace. Either of two common control devices may be used for the automatic shut-off. The oldest and most familiar type is called a "stack control" because it is located in the smoke pipe leading to the chimney — which heating engineers prefer to call a stack. It utilizes "bimetallic strips" like those used in most thermostats, and many other heating-system controls. So, to understand its workings, and the workings of other temperature controls, it helps if you understand the bimetallic strip first. (Heating men usually call it simply a "bimetal.")

The bimetallic strip is simply a thin metal strip made up of two layers, each of a different metal. The metals used are selected because of their wide difference in expansion and contraction rates during temperature changes. Thus, when the strip is heated, one surface of it expands faster than the other, and the strip bends — arching toward the slow-expanding surface. When the strip cools, the performance is reversed. And, when the cooling is continued to lower temperatures, it arches in the opposite direction — because the fast-expanding side (when heated) is also the fast-contracting side when cooled.

To utilize the temperature-responsive qualities of the bimetal, engineers simply harness it to operate switches. When it arches in one direction on cooling, it can make an electrical contact that turns something on. And, when it warms up and arches the other way, it's often used to turn something off. That's how your room thermostat works, perhaps with some refinements. When you set it to turn on your heat when the temperature drops to 70 degrees, you simply move an electrical contact fairly close to the bimetal — so it doesn't have to cool very much before it arches far enough to close the switch. If you set it to turn on the heat at a low 50 degrees, as you might when the house is to be left unoccupied for a while, you are merely moving the electrical contact farther from the bimetal — so it must cool to a lower temperature before it arches enough to work the switch. The dial you turn actually moves the contact.

Playing one bimetal against another. The stack control uses two bimetals to tell whether an oil burner has ignited properly, and to shut it off if it hasn't. The method used is ingenious. Essentially, it's a race between two bimetals.

When the burner starts (on thermostat signal) a tiny electric heater is switched on next to one of the bimetals. This warms it and makes it arch slowly in the direction that will open a switch and shut off the burner. The second bimetal, however, is exposed to the heat that flows into the stack when the burner is in operation. If the burner lights (as it almost always does), the combustion heat in the stack makes the second bimetal tilt the first one so that it can't open the switch and shut off the burner. If the burner doesn't light, there's no tilting, and the electrically heated bimetal shuts everything off. Various other bimetal arrangements are used to produce the same result. Because of the tiny electric heater, they are usually referred to as thermal timing switches. (The tiny heaters are also used to help room thermostats maintain even temperatures. More about this shortly.)

Photoelectric shut-off systems take the place of the thermal type in many new heating units. These utilize a variation of the photocell that moves the pointer in light meters, adjusts automatic camera shutters, and turns on street lights. The cell is mounted so as to focus on the point where the oil flame should appear after ignition. If the flame appears as it should, after the burner starts, the photocell's current holds the electronic power-control switch closed, and the burner keeps on running. If the flame doesn't appear, the photocell's current drops, and the power-control switch opens, shutting off the burner. So there's no danger of fuel-flooding an unlighted furnace. To restart it you must push a "reset" button.

"Limit" controls. After any burner ignites, whether fueled by oil or gas, other safety controls take over. These are the "limit" controls that shut down the burner if the air in the plenum becomes hotter than the manufacturer's design limit, or if the water in a hydronic system exceeds the design temperature or pressure. After the temperature or pressure returns to normal, the limit control switch closes again and the burner restarts. In many hydronic units, controls of this type maintain a constant water temperature, starting and stopping the burner whenever necessary. When the room thermostat calls for heat, it simply turns on the circulating pump to drive the water (already heated to the maintained temperature) through the system. Hence, there's no waiting period for the water to heat up.

Heat-actuated thermostats (on walls) are likely to be found in homes heated by either oil or gas. As mentioned earlier, these use a tiny electric heater to help keep household temperature even. These are designed to reduce or eliminate ups and downs in temperature caused by "overshooting" and "hunting."

They're useful because when your room temperature drops to the figure you've set on your thermostat, your burner starts. Normally, it takes several minutes to heat up the heat exchanger in the furnace or the boiler to the extent that the pump or blower can distribute heat through the house. In the meantime, room temperatures drop a little more. Then heat begins to reach the rooms, and continues to reach them until the bimetal in the thermostat is warmed up enough to shut off the burner. But the thermostat, no matter how good, takes a little time to respond. So the room gets a little warmer than the set temperature before the thermostat acts. Heating men call this "thermostat lag." And even after the thermostat shuts off the burner, the warm air blower or hydronic system pump continues to distribute the leftover heat through the house. (For heating economy, blowers and pumps don't stop running until the central heating unit cools). So, with a simple thermostat, room temperature may climb a bit higher than planned before things start to cool down.

To smooth out the highs and lows, the heat-actuated thermostat "jumps the gun." As soon as the burner starts, one of those tiny electric heaters is switched on next to the bimetal. It warms the bimetal and starts it arching toward the switch shut-off point long before the air in the room is warmed enough to affect it. And it causes the thermostat to shut off the burner earlier in the room warm-up period. But, as a large amount of heat remains in the heating unit, the blower or pump continues to deliver heat long after the burner shut-off. However, instead of sending the room temperature above the level set on the thermostat, the heat-actuated setup comes very close to hitting it on the bull's-eye. Because of the different characteristics of different heating units, the components of this type of system must be matched to each other by the manufacturer.

Natural draft, forced draft, and induced draft are some of the terms you're likely to encounter in the course of selecting a heating unit. *Natural draft,* as the term implies, is the unassisted "pull" of heated air rising up the chimney. *Forced draft* results when the heating unit employs a blower to drive the air through it and up the chimney. The degree to which this is done varies greatly. *Induced draft* is produced when a mechanical means is used to drive air only up the chimney, thus pulling it through the heating unit.

Either of the mechanical systems has the advantage of reducing variations in performance that might otherwise result from wind and outside temperature changes. The blower on a typical gun-type oil burner, for example, drives the combustion air supply into the heating unit, though it may require some natural draft (the pull of hot gases rising in the chimney) for proper performance. An all-out forced draft system pushes enough air through the heating unit to take care of combustion and the exhaust of combustion products up the chimney, too, without requiring aid from natural draft. With a chimney of average height either type performs well. With a low chimney, as in a one-story house, especially where the heating unit is located on the first floor (as in basementless homes) forced or induced draft is often likely to result in improved performance and fuel economy.

The Jet Heet system is an example of forced or "positive" draft, made by Space Conditioning, Inc. The Mark II from the same manufacturer, operates on induced draft, with a blower at the outlet to the flue pipe. There are other units in these categories, both ready-made and designed to fit the situation. The important point is to keep the basic principles in mind and use them where you need them. Heating-supply outlets stock the equipment required and can usually recommend the size and capacity suited to your heating unit.

The pot burner, less common than the gun burner, is often used in large oil-fueled space heaters and in warm air furnaces for small homes and cottages. It gets its name from the fact that the oil fuel burns in what might be called a pot. From the oil-supply line, the oil is led into the base of the combustion pot. No. 1

fuel oil is used (lightest of the distillate fuel oils) instead of the No. 2 (slightly heavier and lower priced.) In the simpler burners a small amount of oil is allowed to flow into the burner pot for starting. Then it is ignited by means of an asbestos igniter, saturated with oil, lighted, and inserted through the starting door of the unit. After the burner pot has been heated by the flame for several minutes, the warmth of the metal shell continues to vaporize the oil, and the flame is established.

Typically, the burner pot is divided into an upper and lower compartment by a ring-shaped baffle about halfway up, with a large hole in the center. (There are many variations, however.) When the oil is first ignited, and when the oil flow is small, the flame is almost entirely below the baffle, and near the oil inlet. As the flow is increased, the flame spreads across the burner pot and hovers just under the baffle. On moderately high fire, some of the flame is inside the upper compartment of the burner pot, the remainder above the pot, rising from the central opening of the top. On full fire the entire flame is above the burner pot. Variations in the form of the burner change the details described to some extent, but the basic principle of operation remains the same.

Most pot burners are of the gravity feed type, though electric pumps are available when the oil must be lifted from the supply tank or barrel to the burner location. To provide an unvarying gravity flow through the regulating valve (adjustable), the oil is led first to a "constant level valve" where the oil level is kept constant by a float valve like the one in your car's carburetor. In some types used for central warm-air heating, the adjustable control valve from the constant level valve is regulated electrically by a thermostat. While this arrangement cannot shut off the burner (as a pilot flame is required to maintain vaporizing temperatures), it can reduce the combustion level to a minimum when heat is not required. Various other versions of the unit include electric ignition forms.

Pot burner draft is important, as downdrafts can blow out the flame, especially while it is on the low-fire or pilot setting. As the draft is reduced by the flame blow-out the burner may not re-ignite, even though hot carbon particles may remain in the burner pot. But oil continues to flow into the burner pot. Under these conditions the burner pot door should *not* be opened until all parts have cooled. Premature opening can admit sufficient air to produce sudden ignition of considerable force, with a resulting flashback through the burner door. So, to play it safe, plan on a chimney height of 25' or more when you use a pot burner, or provide blower-assisted draft. And, if the burner should be blown out by a downdraft, shut off the fuel supply and don't open the burner door until the unit has cooled well below the ignition point. With an adequate chimney, however, the pot burner is a very efficient heating unit with a minimum of moving parts and service points.

HEATING EFFICIENCY. The amount of actual house heat you get from a dollar's worth of fuel depends on how thoroughly your heating unit burns its fuel and how well it distributes it through the house. (Purists have varying interpretations of this, but heat units per dollar are what count when you pay the heating bill.)

Heating's measuring stick is the British thermal unit, abbreviated b.t.u. This is the amount of heat required to raise the temperature of 1 pound of water 1 degree Fahrenheit. This amounts to 252 calories—just about the number in a nice, thick lamb chop.

The rating of heating units is commonly based on their "input" and "output" in b.t.u.'s per hour. The input is the number of b.t.u.'s actually contained in the amount of fuel burned in one hour. The output is the amount of heat delivered by the heating unit to the house during the same hour. This is often given as the number of b.t.u.'s per hour "at the bonnet"—which means at the heat-flow outlet of the unit, in warm-air heating terms. (The amount of heat reaching the rooms is somewhat less because of the normal losses along the ducts. The figure varies with their length and location.)

If the input is 100,000 b.t.u.'s per hour and the output is 80,000, the efficiency is 80 percent, the generally accepted average efficiency for oil and gas units. The difference between the two figures represents the total of such losses as heat escaping up the chimney. But the losses are not clear-cut. Chimney heat, for example, is seldom entirely lost. If the chimney is inside the house its radiated heat contributes considerably to overall heating. In the old days, upstairs bedrooms were often heated solely by radiation from the chimney of a fireplace on the floor below.

To estimate the cost of your heat you need only know the number of b.t.u.'s in a given amount of the fuel and the price of that amount. A gallon of No. 2 fuel oil (the usual residential heating grade), for example, contains an average of 140,000 b.t.u.'s. And you can get the price per gallon from your nearest fuel-oil dealer. Simple arithmetic tells the rest of the story. A cubic foot of natural gas regularly contains about 1000 b.t.u.'s. The cubic foot price is available from your local utility company, but figuring isn't quite as simple as with oil, because gas is sold on a graduated price scale. Rates vary, but the basic pricing method works like this: You pay a monthly minimum charge. Typically, this might be $1.50 for the first 200 cubic feet of gas, or less. The next 1100 cubic feet might cost you 25¢ per hundred cubic feet. The next 1700 might cost 17½¢ per hundred. And all amounts beyond that might cost 12¢ per hundred. (These are actual rates in a suburban area at this writing.)

Electricity is sold by the kilowatt hour (1000 watts used for one hour) and is priced on a somewhat similar sliding scale. For example, in the same area just quoted for gas, the first 450 kilowatt hours (at heating rates) cost 2½¢ per kilowatt hour. For the next 550 the price drops to 1¾¢ per kilowatt hour (kwh), and from thereon, you pay 1½¢ per kwh.

Why some houses cost less to heat. Variations in heating costs do not necessarily reflect on the heating unit or the heating system. The important factor is often the ability of the house to retain the heat. This ability depends on the insulation to block the escape of heat to the outside, and on the difference in temperature between the inside and outside.

Whatever the insulation, it is rated according to the number of b.t.u.'s that can pass through a square foot of it each hour for each degree of temperature difference between the inside and the outside. This heat passage rating is called the insulation's "C value." Typical 2″-thick roof insulation, for example, has a C value of .19 (in fractions, slightly less than ⅕), which means it will pass a little less than ⅕ of a b.t.u. per square foot per hour for each degree of inside-outside temperature difference. So, if the difference is 5 degrees you multiply the C value by 5 and you find your insulation is passing about 1 b.t.u. per hour per square foot.

Types of insulation that have no fixed thickness, like the pellet types you shovel between attic joists, are rated by the "k value." This is like the C value except that it is based on a 1″ thickness of the material.

The "R value" of an insulating material is its resistance to the passage of heat. This is simply the reciprocal of the C or k value. To get it you merely place the figure 1 above the C or k value, making it into a fraction. Thus, if the C value were 4, the R value would be ¼. But for figuring you'd convert it to a decimal, .25.

You need the R values of all the materials in your walls and roof (including lumber and plaster as well as insulation) in order to add up their total insulating effect — called the "U" value. (A partial list is given. More extensive lists are available from insulation manufacturers.) In calm weather, even the layer of still air in contact with the surfaces of the walls (inside and outside) have an insulating value that must be included in the figuring.

To see how it works, you can follow through the typical wall example used by the Insulation Board Institute:

INSULATION	RESISTANCE
Outside wall surface	.17
Wood siding	.85
Insulation board sheathing	2.06
Air space	.97
(Note: a ¾″ air space is minimum that can be used with reflective insulation, and is almost as effective in regular insulating value as 3″ or 4″.)	
Gypsum lath and plaster	.41
Inside surface	.68
Total resistance (R)	5.14

To convert this total to the overall transmission of heat through each square foot of wall per degree of inside–outside temperature difference follow the simple formula: $U = \dfrac{1}{R}$. This gives you $\dfrac{1}{5.14}$, which figures out to .19. So each square foot of your wall will pass .19 b.t.u.'s per square foot per hour for each degree of temperature difference.

Figure out the total area of your outside walls and you can easily determine how many b.t.u.'s are passing through them each hour at any given temperature. You can do the same for your roof, or the ceiling of your upper floor — if the attic roof is uninsulated and the ceiling below it is insulated. Multiply and you can figure the amount of heat lost in a day, month, or year. There are other factors involved in precision heat loss calculations, but this rudimentary method shows clearly how insulation can cut heating costs. And it explains why a poorly insulated house costs much more to heat than a well insulated one.

On the plus side, you and all the other people in your house, plus your lights, pets, and appliances, actually help heat the place, and these are not trivial factors. In marginal heating areas the lights alone provide enough heat to maintain a comfortable temperature in some modern buildings. And if you have ever sweltered in a crowded store after the air conditioning has failed, you can appreciate the heating effect of the folks around you. Remember that each one of them is really a radiator operating at an average temperature of 98.6 degrees. That's why heating and cooling engineers need to know the number of people likely to be in a building before they can make their calculations.

Factors that make heating units more efficient. The object of any combustion type of heating unit is to send the most heat possible into the house while allowing a minimum to escape to the outdoors. Thus, the more heat it transfers to the water in a boiler or to the air around a warm-air heat exchanger, the better. So the more heat-transfer surface the hot gases pass over before they enter the chimney, the more effective the heating job. And, of course, the gases must not pass over these heat transfer surfaces too rapidly.

To accomplish these money-saving ends, heating engineers often use multiple-pass heat exchangers and boilers. These divert the hot gases back and forth or up and down to carry them over heat-exchange surfaces many times the length and width of the unit.

Other means are used to slow the travel of the hot gases up the chimney. One of the simplest and most widely used is the automatic draft regulator. Essentially, this consists of a T connection in the metal smoke pipe from the heating unit to the chimney. The open leg of the T is fitted with a counterbalanced door that tilts on a horizontal hinge pin located a little above the center line of the door. The counterbalance weight attached to the door is adjusted to allow just enough up-chimney draft to carry away the combustion products and to prevent them from drifting out into the house. But if the draft becomes excessive (weather conditions can increase it), the draft regulator's door is sucked inward by the draft pull. This admits cool air from the basement or furnace room, mixing it with the hot chimney gases, and cooling them. Thus cooled, the chimney gases become heavier and rise more slowly, keeping the heat in the heating unit longer for better heat exchange. You can see this type of draft regulator in operation if it is located in the smoke pipe. It is built into some units, however, and can't be seen. This type of regulator is not used with gas-fired units, as a draft diverter (described earlier) must be used instead.

HOW TO REDUCE YOUR HEATING BILL. In any dwelling it is possible to reduce the quantity of heat utilized without seriously affecting comfort and thus reduce heating costs.

The prime factor in heating cost is room temperature. While it is obvious that higher temperatures require more heat, the difference is far greater than readily imagined. Increasing room temperature from 70 to 71 degrees, for example, does not increase fuel costs by 1/70 or 1.4 percent. Actually fuel use is increased by approximately 6 percent. Thus it costs about 30 percent more to hold house temperature at 75 degrees than it does to keep the same house at 70 degrees.

Comfort is only partially dependent on temperature. It is also dependent on the moisture content of the air. Moist air at 70 degrees feels warmer than dry air at 75 degrees. Thus you can operate your home at a lower temperature if you take steps to keep its air moist.

If your home is heated by hot air, you can increase the air's moisture content by installing an automatic humidifier in the heating system. If your home is heated by hot water, containers are available that can be filled with water and hung behind the radiators. If your home is heated by steam, the air is probably moist enough as even the best of air-release valves leak a little steam.

It does pay to turn back the thermostat for short periods of time, even for a few hours. Don't leave the thermostat at 70 degrees all day when no one is home. It can be pushed up for the breakfast period, lowered when everyone leaves for school and work; pushed up for an hour at lunch and down again until supper.

While it is true that if the entire house cools down to 60 degrees, it is going to take a long time to heat the house back up to 70, it is also true that you will actually feel warmer while the temperature is rising than when it has stabilized. This is because the radiators actually "radiate" heat when they are hot. Once the temperature has reached 70 the radiators cool down and there is little radiated heat in the room. By the same token we are more comfortable outdoors at 55 degrees when there is sun than at 65 when it's shady. In a sense, when you raise the temperature for only a short period of time you are warming the people in the house more than the house itself.

The same effect is also the cause of considerable heat waste in many homes. In the evening, when everyone is relaxing, the house seems cooler. The thermostat is then pushed up a few degrees, and while the radiators are hot, everyone feels warmer. When the furnace stops, the rooms feel cooler again and so the thermostat is pushed up once more. Pushing the thermostat up from 70 to 78 for a few hours in the evening can use as much fuel as running the house at 68 all day long.

Another misconception leads people to push the stat up very high in the belief the furnace will produce more heat and so warm the house faster. The result is that no more heat is generated by the furnace, but it keeps running until the preset temperature is reached. Then the house becomes unbearably hot, but lowering the thermostat cannot make up for the fuel wasted.

Additional fuel savings can be achieved by keeping all the doors in the house closed, except for the door at the head of the cellar stairs. Closing the other doors reduces air circulation and cooling. Leaving the cellar door open permits basement heat to drift upstairs where it is wanted.

Turning the heat almost off in unused rooms and turning down the heat in bedrooms to 60 degrees or less are additional heat savers. Once inside your bed you need not be cold even if room temperature is 50 degrees, if you make proper use of blankets. Just remember to place two blankets underneath for every blanket on top. Place the lower blankets beneath the sheet.

Also make certain that both the outside storm windows and the inside windows are closed in all the rooms. This keeps an insulating blanket of air between the two panes.

If you plan on air conditioning later, your ductwork should have larger capacity than for heating alone, as a greater volume of air must be moved for effective cooling. In general, this means that round ducts (often simply called pipe) should be 8″ in diameter for air conditioning rather than 6″, as commonly used for heating alone. (If a system has already been installed using 6″ pipe, it can be adapted to air conditioning by adding more 6″ pipes to carry the added volume of air to and from the rooms.) As the cooled air moving through the pipes and ducts is heavier than the hot air used for heating, it also moves more slowly. So, in adapting an existing system to air conditioning, it is sometimes necessary to also add a higher capacity blower. The performance of the system with the original blower tells the story.

Return-air register locations should also be planned for the system's dual role. In general, return-air inlets for air conditioning should be near the ceiling to draw off the rising warm air. For heating, the return-air ducts are usually in or near the floor to draw off the cold air that tends to settle. To adapt a system to both heating and cooling, return-air ducts may have both floor and ceiling registers. The upper register is opened and the lower one closed in summer, so warm air is drawn off from the ceiling level for recooling. In winter, the upper return register is closed and the lower one opened to pull cool air from the floor for reheating. Various outlet locations are satisfactory if deflector or diffuser type registers are used at the outlets. These mix the outcoming air (either heated or cooled) thoroughly with the room air, eliminating stratification or "layering" of hot or cold air. Details of good and bad combinations of inlet and outlet locations are given in Chapter 12.

10 How to Install a Central Heating Unit

THE FIRST step in installing a central heating unit is *detailed planning.* You must decide on the type of system, the type of unit, and the specific method of heat distribution. The house itself is often a major factor in your decision, as structural features frequently give one heating arrangement an advantage over another.

FORCED WARM-AIR HEATING FURNACES. There are four basic types of furnaces available for forced warm-air heating, in which a blower powered by an electric motor sends warm air at high velocity through ducts running throughout the house.

The **highboy furnace** is a good choice for first-floor installation when the heating ducts are to be run above the unit, as through the attic to room registers. The warm air comes out of the top. You make a plenum to fit the warm-air outlet (or have one made by a sheet-metal worker) with openings for the ducts or other means of heat distribution. The return air enters the unit through the sides or front near the bottom. In many cases no return ducts are required—as when the return air can flow to the furnace location along a hallway.

The **counterflow furnace** is like the highboy in size and shape, but the blower is mounted so that the warm air comes out of the bottom of the unit while return air enters through the top. You use this one where the heating-supply ducts must be run under the floor. The return air may be brought back through ducts or simply through a hallway, as mentioned previously, if the house design makes this feasible.

The **lowboy** is not as tall as either of the other types, but it's slightly deeper from front to back. In its usual form it has both its warm-air outlet and its return-air inlet at the top. It's the type to use in a basement or partial basement where all ducts are to be above the furnace.

The **horizontal furnace** is like a highboy laid on its side. Warm air comes out of one end, return air enters through the other. Use this one in very shallow crawl spaces or attics where all ductwork can be run at approximately the same level as the furnace.

Adaptations may often be made easily. Ready-made units are available—for example, to convert some highboy types for top connection of both warm-air supply ducts and return-air ducts.

DUCT SYSTEMS. There are two types of duct systems available for forced warm-air heating units, the perimeter system (also called the radial system when its ducts radiate outward from the furnace) and the extended plenum system.

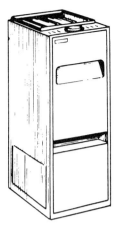

Highboy forced warm-air furnace usually takes return air in back, discharges warm air from top.

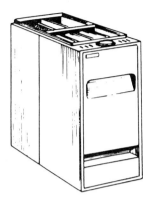

Lowboy usually takes return air in through top, also discharges warm air through top.

Counterflow is similar to highboy, but takes return air in through top, discharges warm air from bottom.

Horizontal furnace takes return air in one end, discharges warm air through the other end.

Each has certain advantages and drawbacks, and each is effective in particular situations.

The perimeter system is the usual one used with modern forced warm-air heating. The term simply means that the warm-air outlets are located around the perimeter of the house—along the outside walls of the rooms. (As described in Chapter 10, the old gravity hot-air systems had their heating outlets along *inside* walls.) There are numerous in-between arrangements, however, with assorted locations of outlet and return air registers, so the term is sometimes applied loosely. But when the house structure permits, you'll usually get the best heating

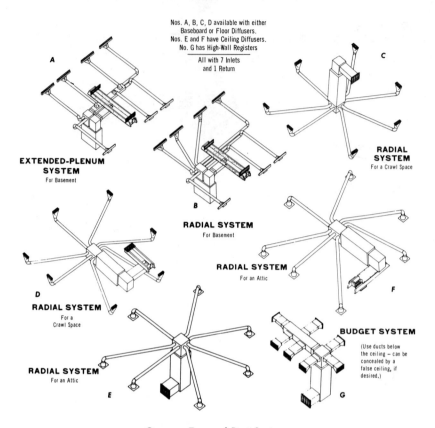

Nos. A, B, C, D available with either
Baseboard or Floor Diffusers.
Nos. E and F have Ceiling Diffusers.
No. G has High-Wall Registers

All with 7 Inlets
and 1 Return

A

C

**EXTENDED-PLENUM
SYSTEM**
For Basement

**RADIAL
SYSTEM**
For a Crawl Space

B

RADIAL SYSTEM
For Basement

RADIAL SYSTEM
For an Attic

F

D

RADIAL SYSTEM
For a
Crawl Space

BUDGET SYSTEM

(Use ducts below
the ceiling — can be
concealed by a
false ceiling, if
desired.)

RADIAL SYSTEM
For an Attic

E

G

Common Types of Duct Systems

results with a true perimeter setup. The return air may be led back through inside wall registers or through the house itself, as mentioned earlier. More about this shortly.

Where space permits, as in the basement or attic, round ducts (like stovepipe) are the easiest to use. In their commonest form, one end of each section of duct is crimped (corrugated), the other plain. The crimped end of one section fits inside the plain end of the next section, forming a simple, airtight joint.

Each run of warm-air supply duct begins at the plenum with a "starting collar" that locks into a duct-size hole cut in the plenum. The duct run can be led around turns of any angle with adjustable 90-degree elbows. These are similar to stovepipe elbows. They're usually made in four sections joined by sliding seams. By rotating one or more of the angled sections around adjacent sections the overall angle of the elbow can be changed. The sharpest turn possible is 90 degrees. For smaller angles, just slip the sections around as required. Where ductwork must run upward through walls, as from one floor to another, a "wall stack" is used. This is simply a duct with rectangular cross section sized to fit between the studs (posts) inside the wall. Similar duct is used for other in-the-structure runs, as between ceiling joists. (Large plumbing and heating-supply dealers and mail-order houses sell 6″ and 8″ round pipe duct and certain widely used rectangular duct, also fittings to go with both types. Tinsmiths can make what you can't buy ready-made, but the price will be higher.)

The furnace location and the layout of the house determine the routes your ductwork must take and the best basic method of heat distribution.

Radial ducting is the easiest and most economical to install — and it's feasible in most houses. Usually it calls for round pipe duct only. In general, each run of pipe extends outward from the furnace plenum to the warm-air register at the perimeter of the house. So, although elbows may be required here and there along some runs, the overall layout is something like the spokes in a wheel.

The extended plenum system is the other popular arrangement. Here a very large rectangular duct (almost as large as the plenum) extends in a straight line from the plenum. Round pipe ducts branch off from the extended plenum to the wall registers. This heat distribution method comes in handy where the basement is to be finished for use as a family room or other living space. The extended plenum, for example, often runs at right angles to the basement ceiling joists, and when properly painted or covered, resembles a heavy beam or girder. The branch pipes can, in most cases, be mounted between the joists so as not to interfere with the installation of a ceiling surface. Another advantage of the extended plenum system is the reduced resistance to the flow of air over a long run.

HYDRONIC SYSTEM FURNACES. Heating units for this system are usually smaller than those for the warm-air system, as boilers can be smaller than air heat exchangers of the same capacity, but its main advantage is that it can supply hot water for faucets year round, even though the heating system itself is not in use. The faucet hot water, however, does not come from the boiler, but is heated in a coil located inside the boiler. The burner (whatever the fuel) operates whenever necessary to maintain hot-water temperature in the boiler. The boiler water

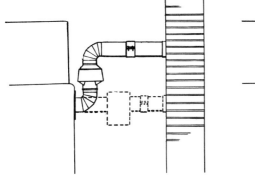
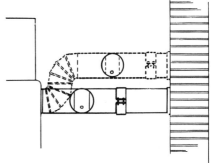

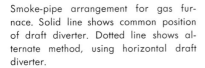

Smoke-pipe arrangement for gas furnace. Solid line shows common position of draft diverter. Dotted line shows alternate method, using horizontal draft diverter.

Usual oil smoke-pipe arrangement is shown by solid line. Dotted line shows alternate setup. Note that draft adjuster is mounted in same position in both arrangements.

heat is transferred through the built-in coil to the water for the faucets. And the transfer is so efficient that many types can be used without a hot-water storage tank. In summer, the pump that circulates the hot water through the heating system does not operate. So, even though the boiler contents are kept hot, the heat does not reach the rooms. But it reaches the hot-water heating coil of the faucet supply. As this requires very little heat, fuel costs are on a par with those of independent hot-water heaters.

Hot-water heating pipes in modern hydronic systems are invariably run around the outside walls of the rooms, though some of the runs must usually be led across the house through inside walls in order to link the various parts of the system together. Although watertightness requires more exacting pipe-joint work than in duct connections, the carpentry involved in leading hot-water heating pipes through the house is simpler than that required for duct work. The diagrams in Chapter 10 show the basic layouts for the common hydronic systems, including the *series loop, one-pipe system, zone system,* and *two-pipe system.* The basic pipe, fittings, and methods of connection, are the same as for your water-supply plumbing.

The capacity of the unit, in terms of its output in b.t.u.'s, must be selected to suit the house. In general, it must be enough to at least keep up with the heat loss in b.t.u.'s through the outer shell of the house. This, in turn, depends on the "design temperature" of the locality—how cold it's likely to be outside during the heating season. This does *not* mean the coldest weather recorded for the area, but for heating purposes, the lowest temperature reached during 97½ percent of the heating season. The small difference between this figure and 100 percent means that a heating system planned according to the locality's listed outside design temperature will be able to maintain the desired temperature level in the house during all but a few brief periods of extreme cold. These often amount to only a few hours during the winter, and are likely to pass unnoticed because the variation of inside temperature is minor.

The other factor is the inside temperature you want to maintain which might be called the inside design temperature. Your heating calculations must be based on sufficient heating capacity to maintain the difference between the outside design temperature and the desired inside temperature. As explained in Chapter 10, the greater the temperature difference the more heat passes through the walls and roof, and the more heat must be supplied to replace the loss. The insulating values of various building and insulating materials listed in Chapter 10 will enable you to make a rough estimate of your heat loss. Another aid to making an educated guess, once you have figured the probable heat loss of the walls, is the percentage of heat lost through the various parts of a house. A study of typical two-story homes (without insulation) reveals relative heat losses in winter as follows:

Walls	32.9%
Ceilings and roof	22.2
Floors	.3
Glass and doors	29.9
Infiltration	14.7
	100%

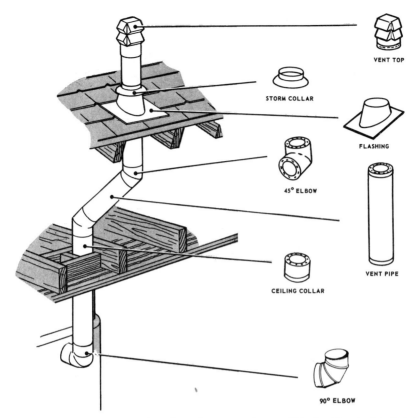

VENT TOP

STORM COLLAR

FLASHING

45° ELBOW

VENT PIPE

CEILING COLLAR

90° ELBOW

How prefab vent pipe can be used in place of masonry chimney. This simplifies the job, but pipe must be a type approved by the code.

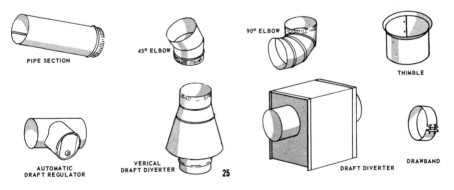

PIPE SECTION

45° ELBOW

90° ELBOW

THIMBLE

AUTOMATIC DRAFT REGULATOR

VERICAL DRAFT DIVERTER

25

DRAFT DIVERTER

DRAWBAND

Smoke pipe and fittings commonly used in run from heating unit to chimney.

The walls show a greater loss than the ceilings and roof simply because they have greater area. The *rate* of heat loss *per square foot*, however, is greatest at ceiling and roof, as warm air rises and creates a greater inside-outside temperature difference there. For that reason, a heating system that provides adequate air circulation to mix warm and cool air, is more efficient. Constant air circulation (abbreviated CAC) with the furnace's circulating blower running at all times is often used to produce this effect. The cost of constantly operating a typical warm-air furnace circulating blower is about the same as that of keeping a reading lamp lighted.

Your heating loss estimates, however, are subject to many variables – the speed of the wind, the exact width of cracks around windows and doors, the rate of infiltration through wall materials, themselves, etc. – and cannot be accurately computed without extensive checking. If you are buying your heating unit or a complete system from a mail-order firm (also some manufacturers) the company may simplify the calculating job for you by providing a heating-load chart. You simply fill in the blanks dealing with the wall and ceiling areas, and other pertinent factors, send it to the company, and have your heat loss (and required heating unit capacity) computed by experts. Unit capacities in b.t.u.'s generally range from around 75,000 per hour output for small homes to around 200,000 for larger ones.

THE FUEL SUPPLY. If your heating unit is gas-fired, the utility company, the local codes, or both are likely to require that fuel piping be installed professionally. The completed installation of the unit and its fuel system, with either gas or oil fuel, must usually be checked over by a local inspector before being put into operation. For this reason, all local regulations regarding heating systems should be known to you before you start work on the job. Because of special requirements in some areas, in fact, it's wise to become familiar with the local regulations before buying your unit. Then, if any special safety device is required, you can order your unit so equipped.

As fuel oil will not burn unless vaporized or in a wick, the supply tank can be either indoors or outside. The most widely used tank is the 275-gallon indoor model. This is 64" long by 42" by 27", and may be installed with either the 42" or 27" dimension vertical. However, because the connecting openings are in different locations for the different positions, the tank must be bought specifically for the desired position. In most basement installations the tank is installed with the 42" dimension vertical. It is installed "flat" with the 42" dimension horizontal when it must be fitted into a shallow crawl space. This position is also preferred when it is to be used with a burner that receives its fuel through a "constant level valve." This is a float-valve device that regulates the flow of oil to the burner, as in pot burners and sleeve types. (For details see Chapter 14.)

Filler and vent pipes leading outdoors from the tank should run as directly as possible. Check your local regulations before installing them, as codes usually specify the sizes to be used. If there is no code, use a filler pipe not less than 2" in diameter. The best type of pipe to use (among commonly available forms) is black iron with malleable iron fittings.

In making connections do not dip the pipe threads into pipe-sealing compound, as this can result in some of the compound getting into the tank. It may then be carried through the feed lines to the burner and clog strainers or nozzles. Instead, screw the pipe into the fitting about two threads, then coat the male thread with the compound, using a small paint brush. Then finish tightening the connection. Use *only* oil-line compound or teflon tape for this purpose. Ordinary pipe compound (for water and drain lines) will be dissolved by fuel oil.

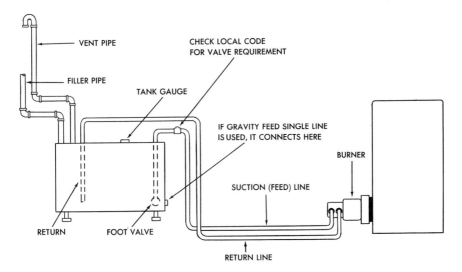

VENT PIPE

FILLER PIPE

TANK GAUGE

CHECK LOCAL CODE
FOR VALVE REQUIREMENT

IF GRAVITY FEED SINGLE LINE
IS USED, IT CONNECTS HERE

BURNER

SUCTION (FEED) LINE

RETURN   FOOT VALVE

RETURN LINE

Usual method of connecting double oil line from tank to burner, and of connecting filler pipe and vent pipe. Double line carried oil to burner pump, providing ample supply for burner nozzle, returning excess to tank. Check local code before rigging tank, as local regulation may require special shut-off valves, fire valves, etc.

Feed lines to the burner in residential installations are almost always of copper tube. The diameter depends on the type of burner and local regulations. An outside diameter of $\frac{1}{2}''$ is common. Regulations usually require that the tubing be run in one continuous length from the tank to the burner, to minimize the possibility of leaks. Some burners require a double line, one tube bringing the oil to the burner, the other returning the excess from the pump to the tank. Others use a single line. Regulations often require that the line be embedded in the masonry floor of basement installations. In any event, it should be protected from mechanical damage.

In some localities "temperature-actuated valves," sometimes called fire valves, are required at the tank end of the line or at both ends. These contain a link of soft metal that melts when exposed to flame, releasing a spring-loaded or weighted valve that snaps shut on release. These are wise safety factors to include in case of fire from any cause. (Even filled oil tanks often survive undamaged through fires resulting from causes unrelated to the heating system.)

Oil feed line connections are usually of the flare-fitting type, and are made as described in Chapter 3. The line itself should be so arranged as to allow some flexing and play at the ends to permit disconnecting the line if necessary. If the burner is removed for cleaning, for instance, the line must be disconnected. If the line leads into the top of the tank, as it commonly does in gun-burner installations, it usually is fitted with a "foot valve" to prevent loss of "prime" from oil draining out of the feed line. The drawing shows a typical plan for a tank-to-burner feed line. Be sure to check your local code for special requirements before you lay out your lines.

Underground tanks may be used where an unusually large oil supply is desired, as where several buildings are supplied on a farm or large estate. Installation of this type of tank should be planned so as to be completed in one day. Otherwise, a sudden rain storm or rising ground water might float the tank upward and break the lines. Once the tank is covered and filled it is secure.

The location of the heating unit is also subject to local regulations, as is the location of the smoke-pipe entrance into the chimney. The distance between the smoke pipe and any combustible material above it, such as basement ceiling joists, is important. If your local regulations do not specify this distance you can play safe by allowing a minimum of 18".

INSTALLING HEATING UNIT AND TANK. Standard 275-gallon tanks for modern heating units are designed to pass through a standard 30" door. Unit weight is also a factor, and should be known before the job begins. Warm-air units of typical design in the 80,000 to 100,000 b.t.u. range are likely to weigh around 200 to 300 pounds. Hot-water boilers of similar capacity, however, may tip the scales at close to 600 pounds. Learn the weight in advance so you can have the necessary help and equipment on hand.

Heavy units are usually laid on one side and slid down the basement stairs on planks, controlled by a block and tackle. Be sure the stairs are strong enough to support the weight and get the manufacturer's recommendations before tipping the unit on its side or back. Many units are specially crated to make this part of the job as simple as possible. A 275-gallon tank weighs about 215 pounds without its pipe legs or other accessories. It's easiest to handle if brought in stripped. Small three-caster "dollies" available from hardware stores ease the job of moving the heavy equipment, and they come in handy later for other jobs. You can also rent heavy-duty dollies.

All heating units should be carefully leveled during installation. With hydronic furnaces be sure to provide pipe unions wherever indicated in the manufacturer's instructions, to facilitate disconnecting the unit if it should be necessary for repairs. Flexible connectors and swing joints (see illustration) simplify connecting pipe to the unit by allowing for minor variations in measurement.

If the unit is to be used on a combustible floor or near combustible walls, be sure you select a model approved for that type of use. In advance of installation, check the manufacturer's installation instructions against your local code. If they are at variance, work out the differences with your local building inspector. Your installation won't otherwise pass his inspection.

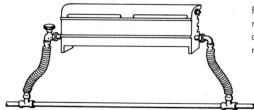

Flexible connectors in hydronic system make job easier by allowing leeway at connecting points. Flexible connectors also reduce vibration in system.

Be sure all automatic controls, expansion tanks, valves, etc., are correctly installed before starting the system for the first time. And follow the manufacturer's starting instructions to the letter. There are important differences in makes and models, and failure to follow the correct starting procedure can sometimes cause not only faulty operation but mechanical damage.

If the unit is going into an old house with a chimney in poor condition, check with your building inspector on approved types of prefabricated chimneys permitted in your locality. In some instances these are actually a more practical and economical answer to the deteriorated chimney problem than repairing the existing chimney.

11 | Installing the Heat Distribution System

The ductwork and air pipes of a warm-air system, and the water piping of a hydronic system, require considerably more time to install than the heating unit itself. But, although extensive, the work is not difficult.

HYDRONIC SYSTEMS. All pipe connections and related work in hydronic installations are handled as described in Chapters 2 and 3. You have the usual choice between copper and steel pipe or tubing. The copper costs more and is hard to get in some areas. But it makes the installation much easier. Steel pipe is available almost everywhere, costs less than copper, and withstands impacts and rough treatment better. But it calls for threading and wrench work, plus a little extra planning to allow for possible disconnecting for heating system repairs at a later date. (Swing joints and unions must be used to facilitate this.) Careful measurements are also necessary to allow for threading distances at fittings. But regardless of the type of pipe or tubing, hydronic systems require less carpentry, as the relatively small water pipe can be led through walls and across floor and ceiling structures more easily than ductwork or air pipes.

WARM-AIR DUCTWORK, once planned, should fit into the house with only minor on-the-spot trimming. If you are buying your complete system from a mail-order house, your ductwork will usually be tailored to fit your house from a dimension chart you have filled out in advance. If you have your plenum and ductwork made locally from your own plans, it should fit equally well – if you have measured correctly. If you are not buying a completely tailored-to-order, mail-order heating system, and have to plan your own runs of duct and air pipe, you'll do well to begin by scouting the larger plumbing and heating supply dealers in your locality.

Ready-made rectangular duct is available in a variety of sizes (your local dealer may or may not stock them all), commonly 8" in their narrow dimension, in widths from 10" to 28". Popular sizes are 10", 14", 18", 24", and 28". These can serve as extended plenums to which round air pipes are connected. Section length is usually 5'. Rectangular duct for "wall stacks" that carry the warm air upward through walls, is also available ready-made. Its thickness is 3¼", to allow leeway inside walls framed with 2-by-4's, which provide a space of about 3⅝" between the wall surfaces. The usual width of the stack duct is 12".

Round air pipe, which may start either directly from the heating unit plenum or from an extended plenum, comes in 6" and 8" diameters (and in larger and smaller ones at some dealers), and in 5' lengths. But you may have to settle for 2' lengths. Although aluminum is now very widely used for ducts and air pipes you'll also find galvanized steel. If you have a choice in your area let your budget be your guide. Both are good. You'll also find fittings to "take off" from the furnace plenum or an extended plenum, and others to interconnect from round air pipe to wall stacks. Other standard fittings cover all other common situations.

As shown in the drawings, outlet and return registers may be either in the

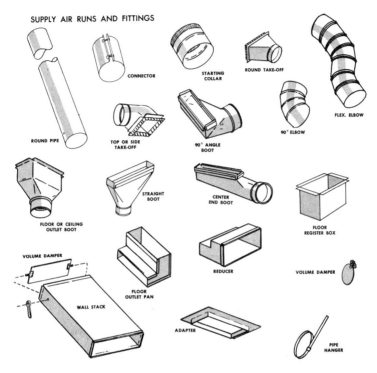

Commonly available pipe and duct fittings. Adjustable round pipe elbows may be joined to make needed zig-zag direction changes in round pipe. Standard size dampers are available for both round pipe and rectangular duct.

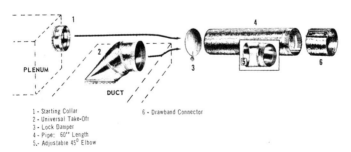

1 - Starting Collar
2 - Universal Take-Off
3 - Lock Damper
4 - Pipe: 60" Length
5 - Adjustable 45° Elbow
6 - Drawband Connector

Starting collar connects round pipe to usual plenum. Universal take-off can be used to connect round pipe either to plenum, extended plenum, or rectangular duct. All fittings and duct sections are designed to fit together, with one small end fitting snugly into large end of next section or fitting. When round pipe must be cut to length, this method won't work, so drawband connector is used.

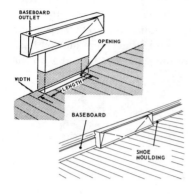

Baseboard outlet is used where outlet cannot be recessed in wall, as in case of solid masonry walls or structural problems. It also eliminates need for cutting through sole plate of wall, making installation easier. Duct opening is made in floor close to wall. Only a section need be cut out behind outlet.

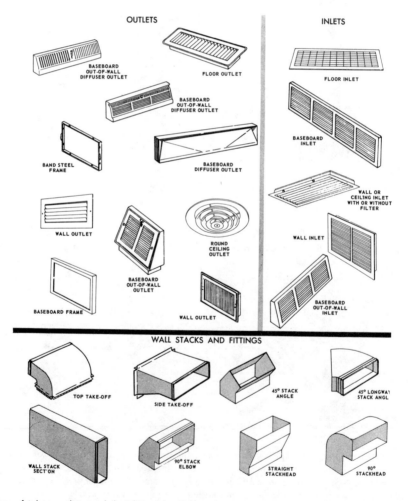

Types of inlets, outlets, and duct fittings leading to them.

floor, in the wall, or just in front of the wall (with backs against the wall surface), depending on the particular structural problem. Often, it's easier to set a register in front of the wall than in it, as it is then not necessary to cut through the 2-by-4 sole plate of the wall framing. Floor registers, while easy to install, may involve rug and carpet problems, and are more likely to collect dust and dirt.

The big advantage of the round air pipes now used on runs from the plenum or extended plenum to first-floor registers and wall stacks (and in return air runs, as well) is their economy and ease of on-the-spot trimming. For example, whether you are working with 5' lengths or 2' lengths, your runs of round pipe are seldom likely to come out in even lengths. So the final section must be trimmed. As each section is made with a plain end (the large diameter end) and a crimped end (designed to fit inside the plain end—and consequently a shade smaller), a cut-to-fit length may not connect in the normal manner. But with round air pipe, a "draw band" connector can be used at the trimmed connection. This is a wrap-around connector with a slide-in wedge section that eliminates any chance of air leakage—and also allows for a reasonable gap between pipe and fitting.

About duct losses. Duct losses are the heat losses that occur along the run of duct and round pipe between the heating unit and the outlet registers in the rooms. They are not true losses because the heat escaping from the ducts remains in the house. But they disturb the balance of a system as they put extra heat where it is not desired and steal heat from areas it is intended to reach. Typically, duct losses dissipate heat near the basement ceiling, where it has little or no comfort value, and starve outlying rooms.

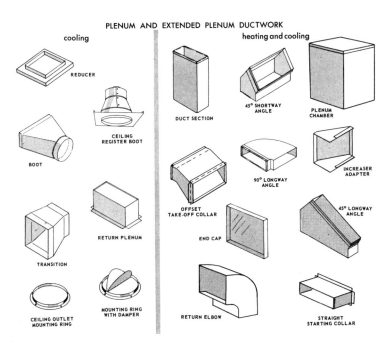

Fittings commonly used with ductwork for heating-and-cooling systems and for cooling alone.

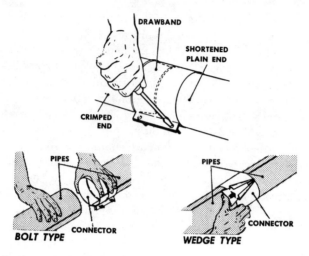

Bolt-type drawband connectors are designed to close snugly and seal pipe when tightened fully. Wedge type is locked and sealed by pointed push-in section.

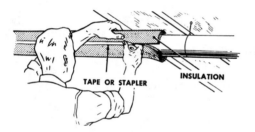

You can buy ready-made insulation for round ducts in various forms. If closure is made with tape use "duct tape" made specially for the purpose and available where you get the insulation.

Duct losses can be cut to negligible proportions by insulating the ducts, extended plenum, and air pipes. For most standard sizes you're likely to find made-to-fit insulation readily available. Do not make the mistake of leaving basement heat supply runs uninsulated with the intention of providing heat for the basement. Heat supplied at ceiling level is likely to leave you with a chilly floor, as heat rises and tends to remain at ceiling level if it originates there. Also, heat from uninsulated ductwork is not mixed with room air by the deflector or mixing type of register. To heat your basement simply include a few outlet registers like those in other rooms. Remember that today's heating units are usually very well insulated. They don't radiate large volumes of heat as did the old coal furnaces.

THE TOOLS FOR THE JOB depend to some extent on the type of heating system. If it is hydronic, one of your greatest aids is usually an electric power drill and a set of drill bits suited to the pipe or tubing sizes you're using. You'll also need the plumbing tools for the type of pipe, as described in Chapter 2. If you

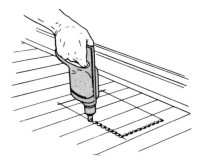

Power drill can be used to make floor register opening by drilling overlapping holes.

You can also cut floor register hole with pistol-grip hacksaw after boring corner holes. Bore holes away from flooring strip edges so you won't hit flooring nails — which can't be seen.

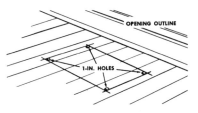

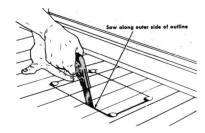

When cutting across flooring strips you can tell by sound if you're hitting flooring nail. Hacksaw will cut through them, or you can steer around them.

Where register hole must be cut all the way up to wall, use the special blade available for the Stanley sabre saw. You can switch to hacksaw type blade when cutting across flooring strips — just in case you hit nails.

are installing a warm-air system you'll need the drill and also a sabre saw if you want to do the job with minimum time and effort.

A sabre saw can make its own starting hole, but the drill is usually better for starting holes because of the close quarters in which this type of work is usually done. To make a hole for a rectangular wall stack, for example, you can drill a hole through the floor at diagonally opposite corners of the rectangle you have penciled on the floor for the stack opening. Starting from these two holes, you can then sabre saw the outline of the opening—cutting first across each end of the outline, then along each side.

Where a cut must be made too close to a wall to permit the use of the sabre saw, you can use a compass saw and do that part of the cut by hand. When cutting through seams between strips of finished flooring, it's wise to use the compass saw rather than the sabre saw—at least as a test. Finished flooring nails are driven through the edges of the strips at an angle, and cannot be seen from the top—often not from the bottom either. If you hear your compass saw grating against a nail stop cutting immediately and change to a metal-cutting blade in your sabre saw. This can cut through the nail and cut far enough into the wood beyond it to permit switching back to the wood cutting blade. As blade changes take only a few seconds with most sabre saws you won't lose much time by this procedure, and you'll avoid blade breakage.

Cutting through plaster or gympsum wallboard can dull a compass saw or other wood saw quickly. So, when you are faced with this job, use a pistol-

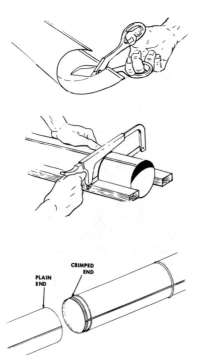

Easiest way to cut most round duct is like this, before edges are snapped together to make pipe.

Where pipe must be fitted into system, then cut to fit, use cleats on board to keep it from rolling and use hacksaw to cut.

Round pipe duct sections fit together like this, with crimped end pushed firmly into plain end of other section. Air flow is always as shown by arrows—don't set pipes the other way.

handled hacksaw. This cuts so fast that a power saw wouldn't gain you enough time to make the blade-breaking possibilities worthwhile. And one or two of the replaceable hacksaw blades will usually carry through the entire installation job.

To cut duct, air pipe, or sheet metal use either a fine-toothed hacksaw, a sabre saw with a fine-toothed metal-cutting blade, or metal snips like Stanley's No. 1543 S aviation snips. In general, if you are cutting round air pipe before it is closed along the seam (it's open when you buy it), the easiest way to do the job is with the snips, as shown in the drawing. However, the snips will squeeze the seam edge closed so the edges won't snap together for closing. You can take care of this quickly by prying open the squeezed-shut section with a thin-bladed screwdriver.

If you use a hand hacksaw to cut air pipe after it has been closed at the seam, block the pipe between strips of wood, as shown, to keep it from rolling when the hacksaw work begins. If you are using a metal-cutting sabre-saw blade on aluminum, you can usually make the starting slit by pushing a penknife blade through the soft metal on the cutting line. (But be careful not to let the blade close on your fingers.) With tougher metal you can usually start the sabre saw from the end of the pipe, then swing it around to make the cut-off, and trim away any protruding metal points after the cut-off is completed. Cutting thin metal is so fast with these saws that a little longer cutting line doesn't make much difference.

REGISTERS, DAMPERS, AND VALVES designed to regulate the flow of heat in warm-air or hydronic systems must be installed with care. Zone controls and manually operated adjusting units frequently must be mounted facing in a specific direction in relation to the direction of flow. Mounted backward, they may not work. Registers, too, usually have a correct "top" and "bottom" when wall-mounted, or a correct inner and outer side when floor-mounted. If mounted in the wrong position, they may deflect the warm air in the wrong direction. And when some registers are mounted correctly, others incorrectly, it is almost impossible to adjust the heat flow correctly. A lever movement that turns one register on turns another off.

BASEBOARD RADIATORS installed in an existing house that is not insulated must have insulation installed in the walls directly behind them. This should extend upward at least 12″ above the floor. To install it, remove the baseboards and cut out a section of the wall from the floor up to a height slightly greater than the thickness of the insulation to be used. A wall opening about 3″ or 4″ high along the base of the wall is usually enough. The insulation (blanket type) can then be pushed up into the wall between studs. When the insulation is in place the opening should be closed with aluminum foil. The housing of the baseboard radiator completes the job.

RADIATING ELEMENTS may be either of the finned tube type or sectional cast iron. If finned copper is used, through-wall connections are made with a

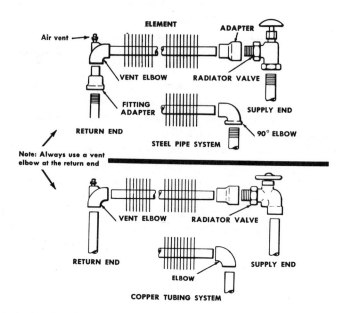

Most finned-tube baseboard radiating units are connected to threaded steel pipe as shown at top. Radiator valve is used at inlet end in one-piece or two-pipe systems. Elbow is shown in small lower drawing, as used in series loop system when valve isn't used. Place air vent to bleed air from radiator. When copper tube is used, radiating units are connected as in bottom drawing. Small lower drawing is for series loop without valve.

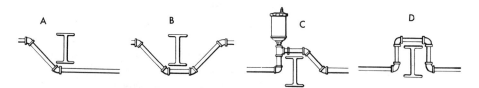

When heating pipe changes levels from under beam to run higher up, use 45-degree elbows (A). Where pipe must be run under beam and up again, use 45-degree elbows (B). Where pipe must run over a beam or girder, an air vent is needed to bleed off trapped air (C). This system (D) has relatively high resistance to flow, also traps air. Avoid it.

Where branch pipes to radiators run between joists, lead them from main in this manner. They should rise initially from main at about 45 degrees to 45-degree elbow, then slope slightly upward to swing joints under radiator, if you're using threaded pipe. Use union in one branch if radiation unit can be rotated on its horizontal axis for disconnecting. Otherwise use union in both branches.

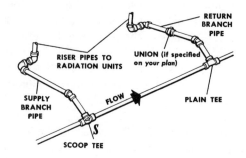

144

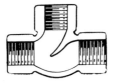

Scoop T, also called forced-flow T, may be used either at inlet or outlet connections from main to radiator, in cases where maximum flow is needed in both branches. When used at inlet connection, scoop should face into water flow, to steer part of water up into radiator. When used at outlet connection, scoop should face in opposite direction.

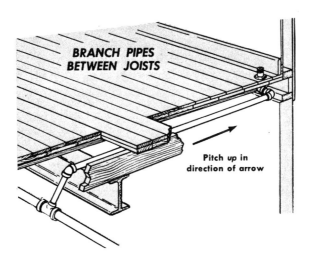

Where branch pipe from main to radiator runs above main, between joists, do it this way with 45-degree elbow at first change of direction after main.

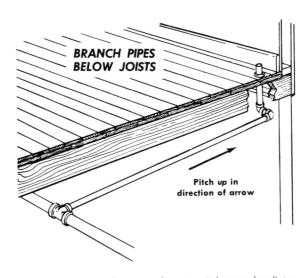

Where branch pipe runs below joists, be sure to have up-pitch toward radiator.

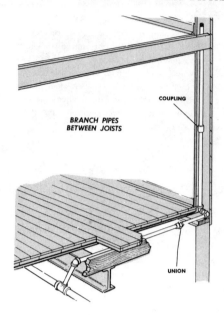

When using threaded pipe up through wall from basement, it's usually easier to use two sections connected by coupling, as basement ceiling is usually lower than upper room ceiling, making it difficult to stand upright. Coupling lets you push first section up until bottom end is at basement ceiling, then screw second into coupling.

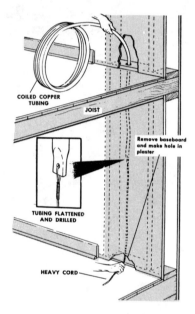

Soft copper tubing often simplifies the job, as it can be snaked through vertical walls like this. At ends, it can be connected to regular tube fittings.

coupling on one side, and a section of tube through the wall into the enlarged end of the tube of the finned section in the next room. The connections are soldered as described in Chapter 3.

There's a trouble-avoiding trick in boring the hole for the pipe to pass through. First measure the height from the floor and the distance from the wall behind it, as specified for the brand of radiator you are using, and mark the point on the wall. Then use an electrician's "feeler bit" to bore a small hole through the wall about half an inch away from the correct point. This small hole through the wall lets you know definitely where the larger pipe hole will come through, and whether any obstacle is in the way. (Inner-wall structural members or pipes sometimes require detouring the pipe, too.) If all's well, as it usually is, the small feeler hole won't interfere with centering the bit for the larger hole, as the small hole is off center. Through-the-wall holes should allow about 1/8" clearance around the pipe or tube. This not only makes installation easier, but prevents

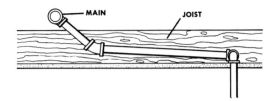

Where boiler is on first floor and must supply radiators in basement, connections from main to radiator are made as usual, but 45-degree initial angle is down instead of up.

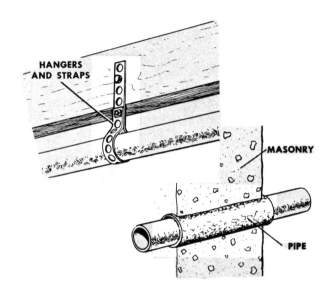

Pipe strap is the simplest answer to pipe support from joists or girders. Where pipe must pass through a masonry wall (sometimes between basement and attached garage) make the hole larger than the pipe and fill in with insulating material like fiberglass.

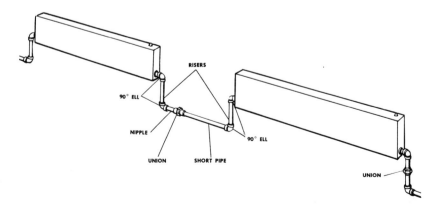

Where radiation units can't be connected through wall from one room to the next, do it this way. Use union wherever there's a possible future need to disconnect work.

creaking noises with expansion and contraction. The clearance can be tucked up lightly with a little mineral wool after the job is complete.

When finned tube must be cut to shorten it, take the length off the plain end of the tube, not the enlarged end. A hacksaw does the job easily. To remove a few fins from the cut end to allow for the connecting fitting, cut the fins inward diagonally with a hacksaw or snips, then twist them off. As cast-iron units are available in assorted lengths, usually made up of factory-assembled 18″ or 24″ sections, you don't have to cut them. But you may have to join sections together. For this you usually need a special assembly tool. Ask about this where you buy your radiators. In all work with either type of radiation element, follow the instructions of the manufacturer. If a data sheet isn't available from the dealer, get it direct from the manufacturer.

If you use threaded pipe instead of copper tube in your heating system, swing joints, as shown in the drawing, make radiator-to-main connections a lot easier. And note that a "union elbow," which can be disconnected like a regular pipe union, is used at the radiator. This makes it possible to remove the radiator without cutting the pipe, if it should ever be necessary. Where copper tube is used with iron radiators tube-to-radiator connections are made with a copper to steel adaptor, as shown in the drawing. On threaded connections, teflon tape can be used to provide a thorough seal without risk of the seal working its way into the piping, as can sometimes happen with paint-on compound.

ADDING A HOT-WATER RADIATOR. As previously described, there are four hot-water space-heating systems in general use. Radiators can as easily be added to one system as any other. However, the location of the new-pipe connections to the existing system will differ with the different systems, as will the size of the pipes and radiators.

Adding radiators to a gravity system is accomplished by connecting the ends of the new radiator to the "hot" and "cold" pipes emerging from the boiler. It makes no difference which end of the radiator is connected to which pipe, but if both ends of the new radiator are connected to the same boiler pipe, no water will flow. The hot pipes are always connected to the top or near the top of the boiler; the cold pipes are always connected near the bottom.

Select a radiator of a size equal to that used in a room with as many windows or total window area as the new room you are planning to heat. Needless to say, large radiators release more heat than smaller ones. Smaller pipe will reduce the radiator's heat output, but larger pipe will not increase heating efficiency.

Connect the new radiator where the other radiators are connected, on the so called mains. If you connect the new radiator to the pipelines feeding an existing radiator, the new and the old radiator together will produce little more heat than the old unit working alone.

Adding radiators to a series loop system requires cutting the feed pipe (the pipe connecting one radiator with another or the boiler) anywhere along its length and inserting the new radiator(s) into the loop.

Problems arise when the new radiator is not close to the feed pipe and the new pipes connecting the new radiator are long. Bear in mind that the feed pipes also act as radiators and therefore must be insulated, if heat is not to be wasted. Also, remember that the new radiator and its feed pipes will slow down the rate of water circulation. The longer the added-on pipes and the smaller the radiator in cross section, the greater the overall reduction in heat flow and heat transference from the boiler to all the radiators. Therefore the new radiators and their pipes must be at least as large or larger than the old. Should you want less heat in the new radiator, use a jumper pipe across the two pipes feeding the new radiator. Installing a gate valve in the jumper will control the quantity of water flowing through the new radiator and thus its heat output. When the valve is open, little water flows; when closed, all the water flows through the new radiator.

Connecting a radiator to a one-pipe system consists of installing two T's, convenient to the new radiator, in the pipe loop, which is connected to the top and bottom of the boiler, and to which the other radiators are similarly connected. One T is standard, the other must be a scoop T. The scoop T must be first in line with the direction of water flow, which is always from the top of the boiler, around the pipe loop, and back to the bottom or near the bottom of the boiler.

Connecting a radiator to a two-pipe system is similar to the procedure outlined for connecting to a gravity system. The pair of pipes connecting each radiator must go all the way back to the boiler. If there is no room at the boiler to add the new pipes, use a pair of T's and short nipples to make new connections to an existing pair of radiator feed lines.

Black pipe, which is nongalvanized steel pipe, is normally used for all hot-water heating pipes. It is a little cheaper than galvanized. All joints are made by threading the pipe and standard black-pipe fittings are used. Valves are brass. Galvanic corrosion is negligible.

Copper tubing can be used, and where it can be snaked through a wall to avoid cutting and patching, copper may be worth its price, but in the diameters necessary for heating, copper is very costly. Use adapters between the tubing and the steel.

ADDING A STEAM RADIATOR. There are two basic steam-heating systems in general use. One is the single-pipe system, which is illustrated. The other, not as common, has two pipes and is more efficient.

To add a radiator to a one-pipe system, connect a single pipe to the steam main and leading it to one end of the new radiator through a radiator valve. Connect an air-release valve to the drilled hole on the farther side of the radiator. With a spirit level atop the radiator, tilt it so the end with the air-valve is an inch higher than the other end, and block it in place. This is done so that water from the condensed steam will run back down the feed pipe.

To add a radiator to a two-pipe system, follow the same steps outlined above. However, a second pipe must be connected to the radiator and that pipe is connected to the water return main. The radiator is not tilted.

12 | Heating System Repairs

TODAY'S automatic heating systems are close to foolproof. Many have been in service for better than twenty years with little more maintenance than a seasonal checkup and an occasional cleaning. But, like any automated equipment, they may shut off at one time or another when one of their sensing devices reacts to a possible malfunction. In many cases the shut-down is due to a condition you can cure quickly, often to something that has nothing to do with the heating unit itself.

CHECK SHUT-OFF SWITCH. First, look for the obvious. If the main shut-off switch to your heating unit is at the head of the cellar stairs, along with a light switch, you may find your heating system isn't operating because somebody flipped the wrong switch and turned it off. Before you call for help, check this point.

CHECK THE FUSE. Look at the fuse that supplies the heating unit in older houses. Today's codes are likely to require a separate fuse for the heating unit. But in pre-code homes the heating unit may share a fuse with some other household wiring. So a short-circuited lamp cord may put your heating system out of commission.

This, and the inadvertantly used shut-off switch, can apply to either gas or oil units—even though most gas units have no combustion air-supply blower. Codes in many areas require a cut-off switch for the heating unit controls regardless of the fuel used.

LOCATION OF THE THERMOSTAT can also be responsible for seemingly mysterious heating troubles. This type of problem sometimes results from a builder's error, sometimes from something done by the homeowner himself after taking possession of the house.

If the house has a single thermostat, and it is located in a room with a fireplace, the operation of the fireplace may create unwanted effects unless you know what to do about it. On a cold night the fireplace may keep the thermostat so cozy and warm that it does not signal for heat even though the rest of the house reaches the cold and chilly level. The best remedy here, is to relocate the thermostat in a nearby room not heated by the fireplace. And *do not* place it on the other side of the fireplace wall.

With this done, the cheery log fire plus the regular heating system may make the fireplace room a little too warm, but you need only close the heating registers in the room or turn off the radiators for the evening. That way, your fireplace heat

150

not only makes the atmosphere more pleasant but serves in a practical manner by actually cutting your regular fuel bill.

Warm walls and sunlight can also create thermostat problems. A thermostat placed in a spot where sunlight strikes it at a particular time of day will be kept warmer than the rest of the room during the sunny period. So the time of day is often a clue to the cause of trouble. While the sun is shining on the thermostat it won't switch on your heating unit when it should, and you're likely to have a chilly house. The best procedure is that of relocating the thermostat. But, in the meantime, you can solve the problem by drawing a curtain or pulling a shade to end the sun-baking.

Cold walls and drafts can have the opposite effect on your thermostat and make your heating unit appear to be the culprit. A thermostat mounted on the inside surface of an outside wall (especially a north wall) is likely to chill down before the air of the room does. So it turns on the heat before the room really needs it and you have an overheated room, probably an overheated house. You can often control this type of ailment by simply turning the thermostat to a lower temperature so it waits longer to respond, but the best bet is relocating it on an inside wall.

A draft striking the thermostat can cause the same effect. If you can eliminate the draft, as by weather-stripping the source, you'll usually solve the problem. If not, move the thermostat. Setting it for a lower temperature won't do the trick, as drafty conditions vary with wind velocity. Hence, you can never be sure of the resulting inside temperature.

PILOT LIGHT FAILURE can shut off your heat in gas-fired units. As described in Chapter 10, the heat of the pilot flame generates electric current (by means of a thermocouple) to keep an electrical valve open to the main gas jet that provides your heat. If the pilot light goes out for any reason (such as a severe chimney downdraft) the valve to the main jet closes so gas cannot escape. With most gas units you can relight the pilot flame yourself, according to the directions printed on a tag or plate attached to the unit. But never try this unless you're sure of what you're doing. To avoid inconvenience, you can call your local utility company in advance and have someone show you the correct way to do the job. Or, if the shut-off occurs before you have a chance to do this, you can get your instructions from the service man who does your relighting for you.

IGNITION TROUBLE in an oil unit can bring about a similar result. Low voltage in the power lines, for example, can sometimes cause the flame-starting spark to lose punch. The flame-sensing element in the unit then shuts off the burner to stop all oil flow. If the low voltage condition is temporary, as it often is, you can put the burner back in operation again merely by pushing the "reset" button on the heating unit. If this does not put the unit back into operation, however, don't keep on pushing it. Call a service man. Pushing the reset button repeatedly can sometimes cause damage.

INSUFFICIENT COMBUSTION AIR is another occasional cause of heating

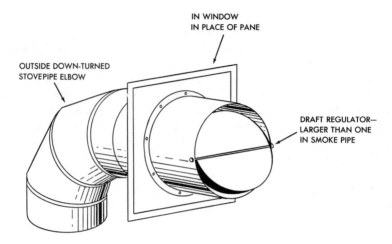

Draft regulator lets air into furnace room without admitting cold air.

difficulties, particularly in newer, tightly sealed homes. In some cases it reveals itself through a slight combustion odor (completely absent under normal conditions) or by other symptoms.

If, for example, your fireplace starts to smoke when your heating unit goes into action, you probably need a greater inlet for combustion air. Your heating unit is actually pulling air down the fireplace chimney for its combustion supply. (Enough air leaks in through the walls and window seams of old houses—not always through new ones.) The cure in any event is extremely simple. Just open a window slightly, somewhere near the heating unit, and leave it that way. You need not open it wide enough to chill the furnace room. And there's a simple way to provide the air only when the heating unit requires it. Just mount a standard draft regulator (larger than the one in your chimney connection) in a furnace room window, as shown in the drawing, and the heating unit will be able to draw in the air it requires. When the burner starts, the up-chimney draft tilts the regulator open, admitting air. When the burner stops, the regulator closes, blocking the entrance of cold air. Use a down-turned stovepipe elbow on the outside of the window to keep out rain and snow. And locate the regulator where cold air can't flow over water pipes inside.

ELECTRIC HEATING TROUBLES. In most cases the result is likely to be localized, as in the case of a cold room caused by a burned-out element, or reduced temperature, as in the case of the failure of a single element in a multi-element central-heating unit. In either event your best bet is to leave the repair to a service man—unless you have installed the system yourself and have the necessary replacement parts and repair know-how. Then, be sure to work only with the current *off*.

ABOUT THE FUEL SUPPLY. Whether you heat with oil, gas, or electricity, your heating trouble may sometimes be caused by a reduced or absent fuel supply. And the evidence is not always apparent. (Many a motorist has run out of gas while his gauge pointed to the full mark.) An unusually heavy demand on gas mains or electric power lines, for example, as in extremely cold weather, can sometimes reduce the heating capacity of the units being supplied. And a delivery error can leave an oil tank with limited reserve supply. To avoid troubles of these types, be sure the mains or power lines to your location are adequate to supply your needs during weather extremes. (Other homeowners using the same fuel in your area are good criteria.) And, if you use oil, arrange for degree-day delivery (see Chapter 10). Dealers calculate this to provide a constant reserve supply, and you can double check your fuel level yourself any time a cold wave is in the offing.

WHEN TO CALL FOR PROFESSIONAL HELP. You're not likely to need professional help often as heating failures are not frequent with modern equipment. But if a check of the simple causes of heating trouble doesn't turn up a cure, don't waste time. Even if you can spot trouble of a more involved nature, and you know how to fix it, you're not likely to have the replacement parts. And, for safety's sake, with any type of heating system, makeshift repairs are not wise. So call for a pro, and give him the make and model of your heating unit and tell him as much as you know about the cause of the trouble. That way, he's likely to arrive with whatever parts he needs to put things back in operation quickly.

STEAM HEAT: PROBLEMS AND CURES. Steam systems are prey to many of the problems attending hot-water systems and a few peculiarly their own. Here are the more common.

Radiator(s) slow to heat. When the trouble afflicts an individual radiator the cause is very often a defective air-release valve on the side of the radiator. One quick way to check is to remove the valve. If the radiator heats rapidly, that is the trouble. Shut off the steam and let the radiator cool; then install a new valve.

When the trouble afflicts a number of radiators the cause may be a defective quick-release valve in the steam main. You can't try the same trick here because the steam will exit from the valve hole and will not reach the radiators. Check this valve by trying a replacement.

Air-release valve emits steam. The valve is defective and must be replaced.

Steam valve emits water. This trouble is usually confined to radiators on the lowest level. It is caused by too much water in the boiler. The cause and cure of this problem is discussed a few paragraphs on.

Noisy radiator. This is usually caused by a radiator that is incorrectly tilted. Water collects inside of the radiator instead of draining off down the feed pipe. Entering steam pounds the water from side to side. Tilt the radiator so the end with the air-release valve is higher. Noise can also be caused by a loose pipe connection and a worn valve in the steam line. The valve stem moves up and down in response to steam pressure. If opening the valve full or closing it tightly stops the noise, the valve is at fault.

Noisy pipes. A pipe that has lost its support and sags will form a "valley" where water collects. Entering steam drives the water out, making noise. If a dropped steam pipe has fallen against a second pipe, the steam pipe will squeak as it expands and contracts with the passage of steam.

Radiator fails to heat. This could be caused by a defective air-release valve, a defective steam valve or a partially opened steam valve. The better systems use gate valves here. Unlike ordinary compression valves, gate valves need upwards of a dozen turns of the handle to be fully opened. Lack of heat in the radiator can also be caused by a very long pipe from the radiator to the steam main and by an open window permitting cold air to chill the steam feed pipe. Close the window and insulate the pipe.

Too much water in the boiler. This usually shows up in water coming out of the radiator air-release valve. High water level is not always visible on the sight glass since the water level will disappear when the water level in the boiler is above the sight glass. The cause is usually failure to regularly "blow down" the automatic feed valve. The valve becomes clogged with rust and sediment and will not fully close. Shut down the system and disassemble the automatic feed after the system is cold. Then clean the valve.

Lack of water in boiler. When all the controls are working properly, a dangerously low level of boiler water will cause the automatic water-feed system to shut the burner off. So, if all controls are on and the burner won't go on, check the water level by examining the sight glass or using the try cocks (a pair of small petcocks at the side of the boiler near the sight glass). With the safety switch on the side of the boiler off, and the boiler cool, open the lower try cock. Water should emerge. If not, the valve (try cock) is plugged up or there is no water above the level of the valve. If there is no water above valve level, the boiler should not be operated; *it is dangerous.*

Lack of water can be due to plugged feed pipes leading to the automatic water-feed mechanism. Disassemble and clear. Lack of water can also be caused by failure to "blow down" the mechanism regularly. Sediment builds up inside the tank and the float remains high, holding the feed valve closed. Disassemble and clean out.

Chronic low water in the boiler can also be caused by very low water pressure and overly high boiler pressure. Under these conditions the water cannot force its way into the boiler except during cool periods. Unless you have your own well and pump system, there isn't much you can do about water pressure, but you should check your steam pressure. It should not be above 5 psi.

For safety and troublefree operation, it is advisable to "blow down" the boiler once a month during the heating season. To do this, open the valve and permit the water and steam to escape into a bucket until the water runs clear. At the same time, it is advisable to keep your eye on the water level and note whether or not the water flows promptly into the boiler after you have let a gallon or so out during blowdown. If the water level does not consistently hug the center of the sight glass, if the water level remains consistently low, find the cause and cure it without too much delay. If the water level has dropped out of sight, leave the cellar immediately, shutting off the furnace as you go. *The boiler may explode.*

Testing room thermostats. You have reason to believe your room stat isn't working, remove its cover and set it at a low temperature. Then shade the device with your hand and set it to a high temperature. Generally you will be able to see the low voltage spark jump across the points exposed to the air or across the points (contacts) inside the little glass vial. If you see a spark, the device is functioning even though it may not be functioning accurately. If you don't see a spark, remove the stat from the wall and use a screwdriver to bridge (short) the two terminals. This is a low-voltage circuit and not dangerous. You should see a small spark and hear the system go into operation. If there is no spark, the trouble is elsewhere. If shorting the terminals starts the equipment, the stat is defective. Try blowing it free of dust and dead moths. If that doesn't help, replace it.

HOT-WATER HEAT: PROBLEMS AND CURES. The problems arising in a hot-water heating system may seem mysterious, but they generally can be solved. There are solid reasons why the system behaves as it does at certain times, though it may not always appear to be so. Here are a few tips to start you thinking.

Radiator fails to get warm, or only a portion of the radiator warms up. The cause could be a pocket of air in the radiator which excludes the heat-carrying water. Open the small (bleed) valve on the side of the radiator. Let the water run until there is no air nor bubbles in the water.

A group of radiators fails to get warm. Trouble is due to a closed or partially closed zone control valve. Trace the radiator feed pipes and make certain all the valves are open.

No radiator heats up, though the burner is going full blast and the circulating pump is working. The trouble is probably a defective thermostat in the main hot-water line. To check the thermostat you have to close the cold-water feed valve that admits water to the boiler. Then you have to drain all the radiators. This is done by opening the petcock at the base of the boiler. Generally this valve has coarse threads on its spout so you can connect a water hose and lead the water out of the cellar as you drain the system. Open the bleed valves on the top floor radiators to speed the process — you can even unscrew a couple.

When the system has drained, remove the large nut on top of the circular housing, a short, fat tube that is on top of the boiler. Reach inside and lift out the thermostat, which looks exactly like an automotive stat, only larger. Note the down side, because that is the way it must be returned. Test the stat by placing it in a pot of water and heating the water. The stat should open at about 150 degrees F. If it doesn't, you can either replace it with a new unit or operate without it. In the latter case you have to make certain to close all the valves in the hot-water heating mains in the spring and reopen them in the fall. Otherwise the water in the boiler will circulate through the system by itself and so heat up the house. The stat also prevents cold water from being circulated by the pump through the boiler, which would chill the domestic hot water. But that generally isn't a problem. If the rest of the equipment is working properly the circulator will be shut down when the boiler becomes cool.

With the stat replaced or the stat housing cover replaced without a new stat, water is admitted to the system and the air is bled out of the radiators and the power is turned on again.

If the stat is satisfactory, the burner is going and there still isn't any heat in any of the radiators, or very little, chances are the circulator is either defective or not running. The water is being heated but is not moving up and through the radiators.

To check the circulator, use a flashlight and *look* at but do not touch the coupling connecting the motor to the pump. If the coupling is turning, the motor is all right but the coupling may be loose. Open the safety switch on the side of the furnace, and then open the furnace switch at the head of the stairs, just to be certain. Now, with your hand, test if the pump shaft is fast to the coupling and in turn fast to the motor shaft. If not, tighten the set screws. If the coupling is damaged, loosen the motor bolts, remove the motor and replace the coupling.

If the coupling is fast to both the pump and motor shafts and you cannot turn the coupling over by hand, the pump shaft is "frozen" to the pump bearing. This happens when you fail to lubricate the shaft with heavy oil at regular intervals. To repair, drain the heating systems as previously discussed and take the pump apart. Sometimes you can loosen the shaft in its bearing by flooding with liquid wrench (a solvent) and then working the shaft loose. If you can do this, remove the shaft from its bearing and rub it down (polish it) with fine emery cloth. If the bearing is too far gone, replace with a new one from a kit. No need to replace the entire pump. Lubricate with SAE 50 oil. Refill the system and bleed the radiators.

If the pump can easily be turned by hand but doesn't turn with the power on, the trouble is electrical. The motor or the controls are defective. Call an experienced oil burner serviceman for help.

Furnace won't start in response to room thermostat, but turns on to provide hot water. There are many possible causes, but one that is easily checked is the setting on the aquastat. This is a small box mounted on the front or side of the boiler. It has a small dial and a control screw beneath the dial. The dial reads degrees, suggesting that the dial setting determines water temperature. It does not. It is a limiting switch; a safety switch. In the summer, when the water may run too hot from the faucet, this control can be turned back to reduce water temperature, which it will do. In the winter, when you need hotter water in the boiler for heating, this control will prevent the water from getting hot enough to permit the circulator to kick it. Thus it is possible for everything to be fine, without the heat ever going on. The usual setting for this control is around 190–200 degrees. The basic purpose of the control is to shut off the burner should the boiler water temperature exceed this figure.

Leaking pressure relief (safety) valve. This is generally caused by lack of air in the expansion tank. The tank should be drained twice a year. When it is filled with water, there is no room for the water in the system to expand with heat. Pressure builds up and the safety valve relieves the excess pressure through the overflow valve.

Another cause is a leak in the immersed, tankless water-heating coil. This coil is connected to the cold-water (and hot-water) pipe. The water is under pressure, almost always higher than the pressure in your hot-water heating system. Therefore when there is a leak in the heating coil, the pressure in the heating system is increased and released through the safety valve.

With the furnace shut off for safety, try turning over the circulating pump by hand. If you cannot turn the coupling, the pump is "frozen." Its shaft ran dry and has actually welded itself (lightly) to the bearing.

Pressure-relief valve opens automatically when water pressure in boiler exceeds preset value. Sometimes dirt will prevent valve from closing properly, so valve leaks even after pressure has returned to normal. Try lifting handle gently and letting a little water out. Valve should be "cracked" (opened) once a year just to make certain it is operative.

The automatic water-feed valve is really a pressure-regulating valve. If it is set at a higher pressure than the relief valve, the latter will open. Loosening the lock nut and screwing the machine screw into the valve increases the pressure of the incoming water. Loosening the screw reduces the pressure.

Water will also drip out of the pressure-relief valve when the pressure on the water feeding the boiler is too high. Feed pressure is controlled by an automatic water-feed valve, which is actually a pressure-reducing valve. Normally the feed valve is adjusted to a pressure lower than the setting on the relief valve. When the feed valve is not operating properly, which can be caused by worn parts and dirty water, or when the valve is set too high, boiler pressure exceeds relief pressure and the safety valve opens.

Automatic water-feed pressure can be varied by loosening the lock nut and turning the projecting screw. Turning the screw into the body of the valve increases pressure; backing out the screw reduces pressure.

13 | Space Heaters

Space heaters are available which operate on electricity, oil, gas, coal, and wood. They can be used to heat garages, outbuildings, or parts of the house that are not adequately heated by the central heating system. Often, too, they are the most economical way to heat an addition to the house that cannot be handled by the existing central unit. Which type you select depends on the amount of heat required, the relative costs of the different fuels in your locality, the availability of a flue, and local regulations.

Certain types of space heaters are banned in some areas. For the most part, this does not mean that the type of space heater is dangerous in itself, but that it has been too often misused by those operating it. Unvented, fuel-burning space heaters (those not connected to a chimney), create a hazard under some conditions by burning up the oxygen in a room and discharging their combustion products into the same room.

ELECTRIC SPACE HEATERS are the easiest to use as they require no flue, do not burn the air's oxygen, and discharge no combustion products. But generally they are capable of supplying only a limited amount of heat at a high operating cost.

Among the portable heaters, the parabolic reflector, the familiar bowl-shaped form, is a handy unit for providing localized radiant heat on a small scale—for example, to keep your feet warm when you work at a bench in a chilly basement. Other portable heaters utilize a blower to combine warm-air circulation with radiation. Portable heaters, for the most part, are limited in capacity to around 1250 watts (about the same as many pop-up toasters) in order to operate from regular wall outlets without blowing a fuse. This wattage allows a margin for lamps or small appliances like radios which may be operating on the same circuit.

Permanent electric heaters include those that are installed in the baseboard, wall, or ceiling, and electric heat-boosters that can be fitted into a central heating duct to increase the flow of heat to the registers in a particular room. These are usually of higher wattage and are commonly rated in b.t.u.'s as well as watts. Their higher wattage, however, may require the installation of an extra circuit to supply the needed current. Many of these operate on 240 volts instead of the regular 120 supplied to most wall outlets. If the heaters are not connected directly to the wiring, special 240-volt outlet receptacles are used. The openings in these outlets are so designed as to prevent the insertion of a cord plug from any 120-volt appliance.

As electric heaters are considered 100 percent efficient, since all of their heat remains in the house, their heating capabilities are relatively easy to figure. Keep in mind, however, that an electric heater mounted in an outside wall may have less than 100 percent efficiency because of heat loss through the wall. The amount of this loss depends on the structure and insulation of the wall. In general, however, you can match the heater to your needs. If temperatures in your area hit a minimum of −30 degrees, you can use a rough rule of the thumb in determining the wattage of the heater or heater required. If you have 4″ of insulation in the ceiling (and ceiling height is under 9′) you can figure on about 1900 watts for each 10′ of outside wall. If, in addition to the ceiling insulation, you have 2″ of insulation in walls and floor, you can probably get by with around 900 watts for each 10′. If the lowest outside temperature is only −20 degrees, you can reduce the wattage to about 1700 and 800 in the two insulating situations mentioned. These figures are intended only as an approximate guide, as the type of house structure, wind velocities, and window conditions can alter the heating efficiency. (Well-fitted windows and storm windows are a help.) In the average situation, the wattages suggested are likely to do the job. Before you buy your units, however, ask if the manufacturer provides a chart for estimating the heater size required for your particular temperature conditions and room size. Many of these, keyed to a particular product, permit surprisingly accurate estimates.

OIL AND GAS HEATERS. Where it is necessary to heat large spaces, you can usually do the job more economically with oil or gas. But a flue or vent is required. (For oil you usually call it a flue; for gas, a vent.) Both types are also available in unvented form if your local code permits their use. These include portables that can be carried from room to room. As mentioned earlier, however, it should be kept in mind that unvented combustion heaters use up the oxygen in a room and discharge combustion gases into it. This can produce a more pronounced effect in a tightly sealed room than in a drafty one. In either case, adequate ventilation should always be provided.

Portable wick space heaters. These old-fashioned heaters can provide 11,000 b.t.u.'s per hour for 6 hours on a mere 3 quarts of No. 1 fuel oil. (This is *not* the same grade of fuel oil used in gun burners, but a lighter form similar to kerosene.) Set at medium burning level the heater can

operate as long as 40 hours on a tankful of fuel. "Wickless burner" portables with twin burners can supply up to 22,000 b.t.u.'s per hour and burn
as long as 24 hours with both burners going. A two-gallon tank holds the
fuel.

The wicks of the round portables usually last through a full heating
season with normal use. They should be trimmed smooth occasionally, according to the manufacturer's instructions.

The wickless burners actually use a small, narrow asbestos strip in the
burner base, called a "kindler." This is lighted after it becomes saturated
with oil shortly after the burner valve is turned on. The resulting flame
heats the metal burner sleeves, after which the vaporized oil burns in gaseous form with a blue flame.

Vented oil space heaters are now usually of the "pot burner" type,
and produce from around 50,000 b.t.u.'s per hour to as much as 85,000 or
more. One of these can heat a cottage, several can heat a full-sized house—
and many houses are still heated by them after more than a quarter century of use. The space-heating pot burner is the same type used in many
central heating units and described in Chapter 10. As mentioned in that
chapter, it should be used *only* with a chimney that supplies good natural
draft, generally one with a height of at least 25' above the location of the
burner. Otherwise select one for which a blower type of draft booster is
available.

Warm-air circulating blowers are available for many cabinet pot-
burner space heaters. These increase the rate of warm-air circulation
through the rooms to be heated. Thermostatic controls are also available,
both for wall mounting and for mounting on the unit itself. Used with a
"constant level valve," as shown in the accompanying photo, these heaters
can be supplied from an outdoor drum or tank. Where this is not possible,
indoor fuel tanks attached to the heater may be used. These are small
enough to carry to an outdoor supply tank for refilling. Humidifiers are
also made to fit the heaters.

Sleeve burner space heaters, now rather scarce, are nevertheless still
in widespread use in outlying country and rural areas. They are among the
most efficient and problem-free heaters. Too, they require only enough
draft to remove the combustion gases and prevent fumes from escaping
into the house, so they can be used with a very short chimney. If you buy a
cottage or summer cabin equipped with one of these burners by all means
keep it. About your only other chance of getting one today is the second-
hand store or the building wrecker's sales yard.

The actual burner consists of a number of concentric, perforated stain-
less steel tubes called "sleeves." The outermost sleeve ranges in diameter
from about 6" to 9", sometimes more. The inner sleeves are sized to allow
about ½" of air space between them. The base of each sleeve rests in a
groove in an oil reservoir (usually cast iron), and the tops are capped by
lift-off steel rings and a center cover. If you acquire a sleeve burner be

Wickless burner for oil space heater is basically a sleeve type with solid outer shell. Outside diameter of burner is usual method of classification, as similar burner sizes produce generally similar heat outputs.

sure all caps and rings are with it, and when you tune it up ready for use be sure they are in their proper places. The accompanying photo shows the components of a common type of sleeve burner and will apply to most of them. But some variation exists. (The wickless burner described earlier is really a sleeve burner with an unperforated outer sleeve.) Oil channels in the oil reservoir at the base of the sleeves are fitted with kindlers of narrow asbestos strip, as in the case of the wickless burner. And similarly, these are lighted to start the burner, after which the oil burns as a vaporized gas. The amount of heat produced by the burner is regulated by a valve that controls the flow of oil to the reservoir at the base of the sleeves. The maximum output is somewhat less than that of the pot burner, size for size, but adequate for much the same type of overall house heating.

All the above-mentioned burners require seasonal cleaning. This is essentially a matter of removing soot and hard carbon from the burner base and of replacing the kindlers in the wickless and sleeve burners. The absence of moving parts eliminates most other problems.

Gas space heaters are, like oil heaters, made in both vented and unvented models, and the same precautions apply to the use of the unvented heaters. An in-between class of gas-fired heaters is also available. This includes the "clay back" heater and the "gas log," both of which are mounted in an existing fireplace. The gas flames are visible and much of the combustion by-product is carried up the chimney. Many gas logs can be lighted electrically by a remote-control switch. Where no flue is available, but where a flue is essential, as in a bedroom, through-the-wall direct vented gas heaters are available that burn outside air only and exhaust their combustion products to the outside as well. Thus, only heat is delivered inside the house, and the inside air is unaffected. Most gas space heaters are available for either bottled or natural gas. But be sure you buy the model adapted specifically for the type of gas to be used. Otherwise serious trouble can result, as bottled gas is likely to contain as much as three times the number of b.t.u.'s per cubic foot as natural gas.

Both oil and gas heaters are available in "floor furnace" form, also. This type is not classed as a true space heater as it is more commonly used as a main source of heat for small houses. It can be used, however, to heat an addition to a house and has the advantage of requiring no actual floor

161

Removing plug-type door in pot-burner case. Off-set spoonlike implement in foreground is burner lighter. Small wad of paper or asbestos saturated with fuel oil is placed in lighter and lighted; then implement is inserted into burner until oil in burner is ignited by it.

Pot type of burner in space heater is lighted through opening in burner case. Plug-type door closes opening.

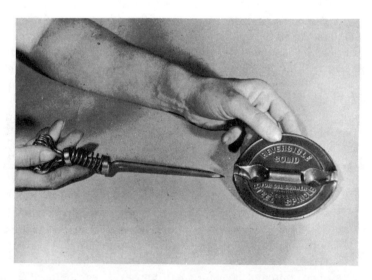

Damper used as smoke pipe of a vented space heater. With pipe unmounted, flat iron disk is held centered inside pipe, near one end. Then pointed handle is pushed through pipe, through gripping slot and tube in center of damper, and out through other side of pipe. Spring holds it locked in place.

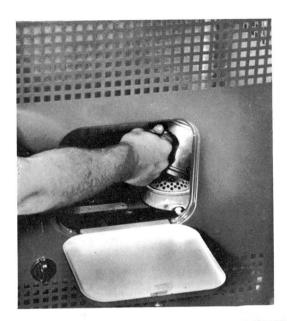

Wickless burner shell is tipped to light it. Match is applied to asbestos kindler in groove in burner base.

"Atmospheric" type of fuel-supply tank for small portable or vented oil space heater. Pin protruding from tank cap is part of valve that is spring-closed when tank is lifted out of oil reservoir below it. When tank is set down in reservoir pin is pushed upward by contact with bottom of reservoir, releasing oil into reservoir.

Porta-Heat midget space heater burns Sterno, heats small room, warms chilly corners of larger rooms. It can also be used for cooking.

Sleeves and top ring and cap of sleeve burner. In 9" diameter, this burner, with no moving parts, can heat an entire cottage. It's noiseless, requires no electricity. If you buy a country retreat with this type of burner, don't throw it away.

space. But the floor grille above it must be left unobstructed. This type is illustrated in Chapter 10. Its b.t.u. output ranges from around 30,000 per hour to about 85,000, so it can heat most anything from a new wing to a small- to medium-sized house. Gas-fueled furnaces are equipped with conventional gas burners and controls. Oil furnaces are available with either pot or gun burners with controls matched to the type of burner.

Wood and coal-burning space heaters are still in production and are still widely used both for their efficiency and their decorative value. They range from the Early American Franklin stove to the familiar pot-belly stove of later times. Most of these require a noncombustible floor for support and a space of 24" to 36" between them and a combustible wall. A layer of brick or hollow tile of comparable thickness is sometimes used under them, atop a combustible floor. But, before installing one of these newly made old-timers check with your local building inspector. Where extra heat is required only occasionally, this type of space heater often serves as an artistic and highly effective aid. And you can count on them for the really tough jobs, like heating an uninsulated outbuilding. How much heat they provide depends mainly on your skill as a stoker and the availability of the fuel. Good, dry firewood generally turns out about 5000 b.t.u.'s per *pound.* And dry coal gives you around 19,000 b.t.u.'s per pound.

14 Fireplaces

A FIREPLACE not only adds a cheery touch to the home but, when in use, supplements the regular heating system. If it is a heat-circulating type it can actually take over a major part of the heating job. It can, for example, supply all the heat required by the room in which it is located. And, if its chimney rises inside the house, it may also provide enough radiant heat to take care of one or more rooms on floors above.

There is nothing tricky about building a fireplace that works properly. It is simply a matter of following time-proven rules. First, don't build a fireplace too big for the room. If you want it to look big you can create the effect of greater size by using a wider facing of masonry around it, or by adding a "Dutch oven" at one side.

You may think of a Dutch Oven as a heavy iron pot with a lid—and you are correct. Early American settlers, however, thought of it as a baking chamber built into the masonry beside a fireplace and fitted with an iron door. Sometimes it was separated from the fire by a metal partition, sometimes by thin stone. In other cases, a small fire chamber was provided beneath it, with a fire door at the front, and an opening into the fireplace at the side—to carry off the smoke of the separate oven fire. The oven fire was often fueled with charcoal for greater heat.

For an average-sized living room an internal width of around 2½' is usually adequate for a fireplace. As a fireplace of this size utilizes shorter logs, it gives you many more cozy evenings from a cord of wood than you'd get from a larger fireplace. Also, it doesn't create the rip-roaring draft for which the big ones are known. The tremendous draft of the old-time log fires pulled such a blast of air through the room that special furniture was devised to protect the occupants from the wind at their backs while they warmed themselves by the hearth. The wing chair, still popular because of its attractive design, got its start that way.

Whatever size fireplace you build, another basic size ratio is important: the area of the fireplace opening should not be greater than 12 times the area of the opening through the flue pipe. Thus, if your flue pipe is a foot square inside (not a standard size, but used to simplify arithmetic) the area of your fireplace opening should not be more than 12 square feet. This might mean a width of 4' and a height of 3'. There's no harm, of course, in having the flue a *little* over-sized. This puts you on the safe side as far as possible fireplace smoking is concerned. But don't go overboard— say, with a 50 percent larger flue area than you need. This just wastes heat up the chimney. And if your chimney is a short one, as is the case in many modern one-story houses, don't expect an over-sized flue pipe to boost your draft. Tests have shown it won't. But a few feet of extra chimney height will help. This doesn't necessarily mean extensive masonry work. An extra length of flue tile bedded in cement at the chimney top, with a few inches of cement around it, often does the trick. Smoothly done, it can top off the existing chimney in attractive style. The reason for a few inches of cement around the extension is that unless the extension has some insulating value in its shell, it chills in the cold winter air and doesn't have much draft-increasing effect.

Keep the 1 to 12 proportions in mind, too, if you have an old house with a smoky fireplace. You may have to climb the roof to measure the flue outlet. If so tie a husky rope around your middle, throw it over the roof peak, and have somebody keep a firm grip on it as you climb the roof. And stay on the opposite side of the peak from your safety man. This way, you're more likely to avoid a serious fall in case you slip. It's a good idea, too, to make the end of the rope fast to something solid—just in case your safety man loses his grip.

If you find your flue is too small for your fireplace opening, it's a lot easier to make the fireplace opening smaller than to tear the whole chimney apart to make the flue bigger. Full details about this shortly.

If you are building you can use the fireplace diagram and chart to

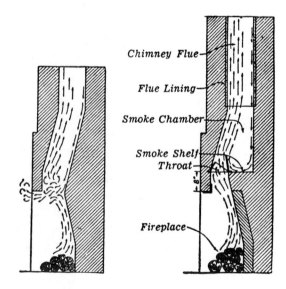

Why fireplaces have a "smoke shelf," and what it does. Old-time fireplaces lacked smoke shelf, and downdrafts flowed into fireplace and puffed smoke into room as at left. Modern fireplaces have shelf, block downdrafts, as at right. Note that damper is mounted on smoke shelf so that, when opened, as necessary when fireplace is in use, damper acts as additional baffle to downdrafts. *Courtesy Donley Bros. Co.*

Chimney Flue

Flue Lining

Smoke Chamber

Smoke Shelf
Throat

Fireplace

establish the dimensions of your fireplace. These guides were developed by the Donley Brothers Co., of 13900 Miles Ave., Cleveland, Ohio, to cover fireplace widths from 2′ to 6′. Dampers and other fireplace equipment you're not likely to make yourself, including Dutch Oven doors and heat-circulating fireplace shells, are available from the same company.

DETAILS OF FIREPLACE DESIGN. If you're more familiar with furnaces than fireplaces, some details of fireplace design may not seem to make sense at first glance.

The opening from the fireplace (through the open damper) to the smoke chamber above it is near the *front* of the fireplace. So it draws the smoke forward, which might seem to invite the smoke to billow out into the room. But it works the other way. Because the up-chimney opening is near the fireplace front any smoke drifting forward toward the room is swept up into the smoke chamber as it tries to pass the suction area of up-chimney opening. If the opening were at the back, the smoke would not have to pass the up-draft on its way forward, and could escape into the room.

Long experience shows the effectiveness of the front-damper. The forward-leaning fireplace back reflects heat downward through the fireplace opening to warm your legs and feet more efficiently. The out-splayed sides also reflect heat into the room instead of simply bouncing it back and forth across the fireplace as squared sides would.

The smoke shelf is there to block downdrafts that might otherwise puff smoke into the room. Note that the damper, which must be open when the fireplace is in use, opens by tilting backward to form an additional barrier to the downdrafts. The right and wrong diagrams show what happens if the smoke shelf is omitted. The ash pit and ash dump don't affect the burning performance of the fireplace, so that may be left out if you want to save work during construction. But they're well worthwhile, as they reduce the chores of removing ashes over the years the fireplace will be used.

The smoke chamber above the fireplace damper is a sort of "cushioning" space that allows smoke to pile up briefly when downdrafts slow the up-chimney flow. And, of course, the chamber also reduces the chance of the downdraft reaching the fireplace proper.

CONSTRUCTING A FIREPLACE. Work on a conventional masonry fireplace begins at the foundation footing, as shown in the diagrams. The hearth in front of the fireplace is cantilevered out from the masonry support of the fireplace. It is *not* supported by the floor framing of the house, as settling of the chimney base and flexing of the floor structure of the house, however slight, could cause cracking of the masonry. The cantilevered hearth, however, must be built with metal reinforcing rods embedded in the concrete to give it structural strength. The rods are available from the same masonry supply outlet as the brick and concrete. (If there's no

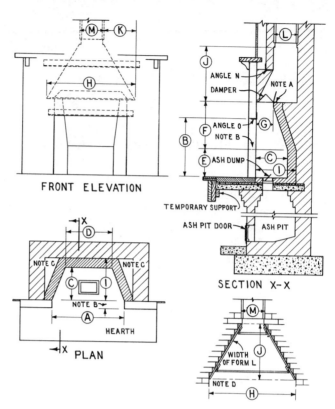

FRONT ELEVATION

SECTION X-X

PLAN

HEARTH

NOTES ON CONSTRUCTION SKETCH AND DIMENSION TABLE

Note A — The back flange of the damper must be protected from intense heat by being fully supported by the masonry. At the same time, the damper should not be built in solidly at the ends but given freedom to expand with heat.

Note B — The drawing indicates the thickness of the brick fireplace front as four inches. However, no definite dimension can be given for this because of the various materials used — marble, stone, tile, etc., all having varying thicknesses.

Note C — The hollow, triangular spaces indicated in the plan, behind the splayed sides of the inner brickwork, should be filled to afford solid backing. If desired to locate a flue in either space, the outside dimensions of the rough brickwork should be increased.

Note D — A good way to build a smoke chamber is to erect a wooden form consisting of two sloping boards at the sides, held apart by spreaders at the top and bottom. Spreaders are nailed upward into cleats as shown. The letters H, M, and J correspond to letters in the elevation and in the Table of Dimensions. The form boards should have the same width as the flue lining. *Courtesy Donley Bros.*

basement, or you're building on a slab foundation, you don't need the cantilevering.)

The supporting base of the fireplace (which contains the ash pit, if you include one) is first built up to a point just far enough below the finished floor level to permit pouring the concrete and surfacing it with whatever you want—flagstone, brick, etc.—so as to end up with the finished hearth flush with the floor. Scrap lumber or plywood can be used to make the form for pouring the cantilevered hearth. But *don't* omit the reinforc-

Table of Fireplace Dimensions

Finished Fireplace Opening							Rough Brickwork and Flue Size			New Flue Sizes**			Round	Old Flue Sizes			Ash-pit Door	Steel Angles*	
A	B	C	D	E	F	G	H	I	J	K	L	M		K	L	M		N	O
24	24	16	11	14	18	8¾	32	20	19	10	8x12		8	11¾	8½x 8½		12x8	A-36	A-36
26	24	16	13	14	18	8¾	34	20	21	11	8x12		8	12¾	8½x 8½		12x8	A-36	A-36
28	24	16	15	14	18	8¾	36	20	21	12	8x12		10	11½	8½x13		12x8	A-36	A-36
30	29	16	17	14	23	8¾	38	20	24	13	12x12		10	12½	8½x13		12x8	A-42	A-36
32	29	16	19	14	23	8¾	40	20	24	14	12x12		10	13½	8½x13		12x8	A-42	A-42
36	29	16	23	14	23	8¾	44	20	27	16	12x12		12	15½	13 x13		12x8	A-48	A-42
40	29	16	27	14	23	8¾	48	20	29	16	12x16		12	17½	13 x13		12x8	A-48	A-48
42	32	16	29	14	26	8¾	50	20	32	17	16x16		12	18½	13 x13		12x8	B-54	A-48
48	32	18	33	14	26	8¾	56	22	37	20	16x16		15	21½	13 x13		12x8	B-60	B-54
54	37	20	37	16	29	13	68	24	45	26	16x16		15	25	13 x18		12x8	B-72	B-60
60	37	22	42	16	29	13	72	27	45	26	16x20		15	27	13 x18		12x8	B-72	B-66
60	40	22	42	16	31	13	72	27	45	26	16x20		18	27	18 x18		12x8	B-72	B-66
72	40	22	54	16	31	13	84	27	56	32	20x20		18	33	18 x18		12x8	C-84	C-84
84	40	24	64	20	28	13	96	29	61	36	20x24		20	36	20 x20		12x8	C-96	C-96
96	40	24	76	20	28	13	108	29	75	42	20x24		22	42	24 x24		12x8	C-108	C-108

NOTES.   *Angle Sizes: A—3x3x³⁄₁₆, B—3½x3x¼, C—5x3½x⁵⁄₁₆.

**New Flue Sizes—Conform to new modular dimensional system. Sizes shown are nominal. Actual size is ½ in. less each dimension.

ing rods. They prevent cracking and they assure you that your hearth can't break off and drop into the cellar, if it's built over a cellar.

The inside surfaces of the fireplace are lined with fire brick set in fire clay, so the rough brickwork must be dimensioned to allow for the lining. The sides and lower portion of the fireplace back, being vertical, make the setting of the firebrick a simple matter. And the forward-sloping upper portion of the back is almost as simple, as the slant is at a relatively slight angle. The more steeply sloped sides of the smoke chamber need not be firebrick-lined unless your local building code requires it.

Once the supporting foundation of the fireplace is built up from ground level, the vertical portions of the sides are erected, as shown, to form a plain, square-cornered masonry box.

Next, the wooden form for the slant-sided smoke chamber is assembled and set in place. (This can be made from scrap lumber.) It's a good idea to make this as wide as the interior exposed brickwork will be. Then you can use ample cement, troweled on to the form, so as to get a com-

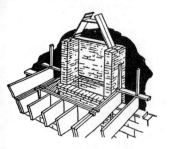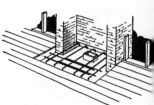

Steps in conventional fireplace construction. Tapered smoke chamber above the fireplace is built on angled wooden form, as shown in left drawing. Base of fireplace extends to footing in basement, as shown in preceding diagram that includes ash pit.

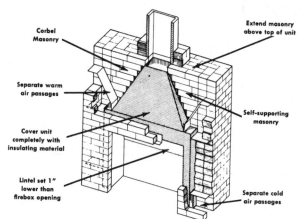

Corbel Masonry

Extend masonry above top of unit

Separate warm air passages

Self-supporting masonry

Cover unit completely with insulating material

Fireplace built around ready-made heat-circulating unit, like this Donley Heatsaver, eliminates chance of dimension errors. Inner metal shell assures correct proportions.

Lintel set 1″ lower than firebox opening

Separate cold air passages

pletely smooth inside surface. It's a good idea, however, to cover the wood with two layers of tar paper to make the form easier to remove after the cement sets. Throughout this stage, keep in mind that the form must be removed. For example, don't leave any little blobs of cement to harden where they might block removal of the form. If you happen to leave a few that you don't discover until the next day, chip them off with a cold chisel as soon as you find them. They'll be firm after a night of hardening, but they won't be rock hard, as cement continues hardening for weeks after it first sets. If any tar paper sticks in place, it will burn off the first time the fireplace is used.

Where firebrick lining must be set at steep angles that might cause it to fall off in case of a flaw in the cement work, use metal "ties" to lock it in place. These are made in various forms and are available from the same sources as your bricks. They're cemented into the rough brickwork and led out between the firebricks to hold them in place. To be sure you know how to use the type you buy, ask about the details where you buy them.

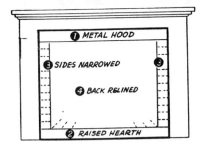

① METAL HOOD

③ SIDES NARROWED ③

④ BACK RELINED

② RAISED HEARTH

If you have a smoky fireplace in an old house because of an undersized flue, here are easy ways to cure the trouble. Try a wide, black, iron hood first. If this has to be too wide to make flue proportions correct, use bricks to raise the hearth — further reducing the fireplace opening — and, if necessary, add a layer of brick to each side.

If you have trouble with smoke or fumes "pouring" from one flue into another and bringing smoke or odors into the house, here are two cures. Left, raise one flue outlet 4″ to 6″ above the other. Right, cap chimney, and build narrow brick walls between flue outlets. Flagstone makes a good chimney cap.

It also helps to support this type of brickwork with temporary wood bracing until the cement sets. A few cut-to-size boards with props wedged in place will do the job.

The damper and the fireplace breast are set in place after the brickwork of the fireplace and the smoke chamber are completed, as shown in the drawings. This is an easy job as the entire front of the structure is wide open at this point. You'll be wise to buy a ready-made damper as you're likely to spend almost as much on materials if you try to make your own. And the ready-made job is likely to last the life of the house. But get a good one. It's not easy to get it out and replace it after the job's done.

When you mount the damper you also install the masonry front of the fireplace on each side, to support the angle iron that holds up the fireplace breast. As the drawings indicate, there are no complicated aspects of this job. The flue lining from the top of the smoke chamber should, of course, be set in place as the chimney brickwork progresses.

Fireplace flues. If two fireplace flues are installed in a single chimney (as from fireplaces in adjoining rooms), or if a fireplace flue and a furnace flue share the same chimney, the flues should *not* end on the same level at the chimney top. Make one at least 4″ to 6″ higher than the other. This eliminates the chance of smoke or fumes from one flue being drawn down the other—as when one fireplace is in use and the other cold. This can make a fireplace smoke even when there's no fire in it. And it can make an unused fireplace belch furnace fumes. It's more likely to happen in modern houses that are more completely air-sealed than the older ones. If a window isn't partially opened in the furnace room or the room where a fireplace is located, the furnace fire or the log fire may pull its combustion air down the unused flue—billowing smoke or fumes into the room. So always provide ventilation where there's a warming fire—especially a furnace fire.

Tip for the furnace room: To avoid chilling the room with cold outside air when it's not required, you can use a standard draft regulator, mounted in place of a basement window pane as shown in the drawing. It should have a down-turned galvanized stovepipe elbow on the outside, to prevent the entry of rain or snow. When the furnace starts, the up-draft in the chimney lowers the air pressure in the furnace room, pulling the balanced flap of the draft regulator open. (It does just about the same thing in regular operation.) This admits the required air to the room. When the furnace stops, the draft regulator closes and keeps cold outside air from getting into the room. This arrangement saves heat and also prevents your central heating unit from drawing its air down your fireplace flue and bringing smoke into your living room.

CIRCULATING FIREPLACES. Air-circulating fireplaces built around ready-made metal shells simplify construction greatly. The brickwork is simply built around the outside of the metal shell, assuring that all angles and proportions are correct. The metal must be left exposed to the fire in order to make the desired air-heating job possible—as in a warm-air furnace. In working with a prefabricated shell of this type, however, always follow the manufacturer's instructions. There can be considerable difference in installation between makes.

FREE-STANDING FIREPLACES. Widely available through lumberyards and building-supply dealers, these provide the warmth and cheery atmosphere of the log fire with considerable reduction in initial cost in many instances.

Ready-to-install fireplace by Vega Industries includes metal fireplace form, smoke dome (smoke chamber), smoke shelf, and damper. Chimney is made in 2' and 3' lengths, with 10" inside diameter. Insulated, Underwriters' tested and approved. Simulated brick top section is also available in gray or white.

Free-standing fireplace with smoke pipe can be connected to existing chimney or to ready-made roof-mounted chimney.

They are commonly sold with prefabricated chimneys. Although the chimneys may be widely approved (after lab testing) in many areas, you'll have to check your local code to be sure the type you want to use is approved in your area. The advantage of the pre-fab chimney lies not only in the cost saving over masonry, but in appearance, as most free-standing units tend toward modern design, and the pre-fab chimneys match the style.

If you have to, you can usually connect a free-standing fireplace smokepipe to any existing masonry chimney that isn't being used, as in the case of old houses that once relied on wood, coal, or oil for cooking. Just be sure the old chimney is in good condition.

No special under-floor support is required for the usual free-standing fireplace, as overall weight is well within the normal support range in most cases.

15 | Air Conditioning

Perhaps without realizing it, most of us are familiar with many of the principles of physics utilized by air conditioners. We know, for example, that when a liquid evaporates it absorbs heat in large quantities, and it takes that heat from whatever happens to be near it. We experience that phenomenon every time the moisture of perspiration evaporates to cool our bodies in hot weather. And if you splash a little rubbing alcohol on your skin it evaporates even faster, and cools you still more.

An air conditioner does its cooling by allowing liquid to evaporate inside it. But it doesn't let the evaporated liquid get away. The evaporation takes place in a cooling coil located *inside* your house, so it absorbs heat from the room. Air blowing over the coil loses its heat, so it comes away much colder than it was before. That's what gives us the refreshingly cool stream of air we want on a hot day.

The heat absorbed by the evaporating refrigerant inside the air conditioner is pumped *outside* to a condenser. The condenser is actually a radiator into which the evaporated refrigerant is pumped under considerable pressure.

If you have ever inflated a bicycle tire with a hand pump you may have noticed that the base of the pump grew hotter as the job progressed. When you compress air or any gas (like the evaporated refrigerant) it gets hotter. The heat that was in it in the first place is still in it, but it's crammed into a smaller space so it is concentrated, more intense. The more you compress the gas the hotter it gets.

But why compress the refrigerant? We want to get the heat *out* of it, and heat travels *only* from a higher temperature area to a lower temperature area. When the evaporated refrigerant comes out of the house on its way to the condenser outside, it is only at approximately the temperature of the air around it. But when we compress it we raise its temperature so that heat flows out of it to the air around the condenser, as the refrigerant is now much hotter than the outside air.

To permit this build-up of pressure in the outdoor condenser, only a tiny tube called a capillary is provided to carry the refrigerant back into

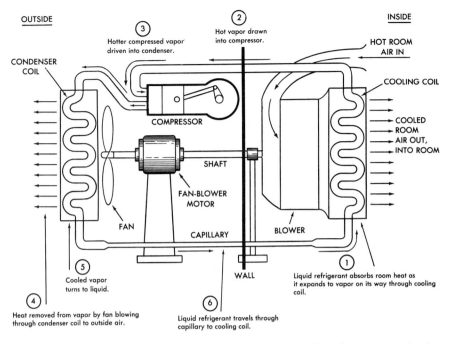

③
Hotter compressed vapor driven into condenser.

②
Hot vapor drawn into compressor.

HOT ROOM
AIR IN

CONDENSER
COIL

COOLING COIL

COMPRESSOR

COOLED
ROOM
AIR OUT,
INTO ROOM

SHAFT

FAN-BLOWER
MOTOR

FAN

CAPILLARY

BLOWER

⑤
Cooled vapor turns to liquid.

WALL

①
Liquid refrigerant absorbs room heat as it expands to vapor on its way through cooling coil.

④
Heat removed from vapor by fan blowing through condenser coil to outside air.

⑥
Liquid refrigerant travels through capillary to cooling coil.

Operation of a typical window air conditioner with cooling coil and condenser in a single, close-coupled housing.

the house. So, with its outflow greatly restricted, the refrigerant remains under constant pressure from the compressor pump. (You might compare the condenser to a tire containing air under high pressure, but having a slow leak.) As a fan blows the outside air over the condenser, cooling the refrigerant inside it, the refrigerant is transformed from a vapor to a liquid again. When it makes its way back into the cooling coil inside the house (passing through the capillary on the way), it is in liquid form. Once in the cooling coil it expands into a gas again as it takes on heat, and the cycle starts all over again.

One of the big advances in the development of air conditioning and refrigeration was the improvement of refrigerants—the substances that are transformed inside the system from the liquid to the gaseous state. They had to be nonpoisonous, noncorrosive, and nonexplosive. One of the earliest refrigerants was ammonia, but it was not entirely satisfactory. Eventually chemists made molecular revisions in carbon tetrachloride—the dry-cleaning chemical that won't burn but causes liver damage if inhaled too much. When they removed two atoms of deadly chlorine and replaced them with two atoms of even deadlier fluorine, they produced a compound with the chemical name dichlorodifluoromethane. It proved to be a safe and efficient refrigerant. Today we know it under the trade name Freon, and it's used for a wide variety of cooling and refrigerating purposes.

HOW BIG A UNIT? The capacity of an air conditioner depends on the size of its cooling and condensing coils, the volume of refrigerant that moves through it, and the power it has to do its job. Therefore your first problem in air conditioning a house is deciding on a unit of the size to do the job.

Engineers and manufacturers have worked out a capacity rating system which conveys some idea of a unit's cooling ability. The system compares the cooling ability of an air conditioner with a specific amount of ice. A 1-ton air conditioner, for example, provides as much cooling effect as the melting of 1 ton of ice each hour. This means it can take 12,000 b.t.u.'s out of the air inside your house in one hour—the same amount a ton of melting ice can eliminate.

Figuring the size of the air conditioner you require is a rather extensive job when you want precise results, but you can use one rough rule of the thumb. If you figure 1 ton of cooling capacity (12,000 b.t.u.'s) for each 500 square feet of floor area, you're likely to come out on the safe side in most areas. But this is just a very broad guide, and it may result in you're buying a bigger unit than you need—sometimes a smaller one.

If considerable outlay is involved you'll be wiser to use the calculating system adopted by the Air Conditioning & Refrigeration Institute. This utilizes a standard form on which you fill in all pertinent information concerning your home and its insulation and surroundings, including wall, roof, and floor areas, and the type of construction and insulation. It even takes into consideration shade trees and types and sizes of windows.

Fold-out forms and tables of Air Conditioning & Refrigeration Institute cover 17″ by 33″ area when opened, form complete calculating system for figuring rating of air-conditioning equipment required for total house cooling. Each set contains all forms, instructions, and twenty-five calculating sheets.

Even the color of your roof is considered, as dark colors absorb more heat than light ones.

Once you fill out the form, you add up the figures the chart prescribes for the different elements and multiply them by specific factors given. This gives you a fairly accurate estimate of your air-conditioning needs. Be sure, however, that you don't merely guess at the important items. Try to get an actual look at the insulation in your walls. You can often do this either in the attic or in the basement. Sometimes the insulation shows inside removable wall panels that afford access to plumbing connections.

Window air conditioners in wide range of capacities can be used on ordinary 115-volt outlets of 15 to 20 amps. Larger models (in 22,000 b.t.u. range) can also be window-mounted and supplied by 115 or 230 current. Typical medium-capacity units can be mounted in less than a minute.

ROOM UNITS. Your choice between room units or a central system depends on your individual needs. From the standpoint of easy installation, the room unit has the edge. Some of these can be carried easily by one person and can actually be installed in a window in less than a minute. They need only be plugged into a wall outlet to be put in operation.

If you prefer not to obstruct window area, you can use the same type, or buy one made for the purpose and mount it in a wall opening cut and framed to fit. Just be sure the air conditioner is located at least as far above the floor as it would be in a window. Otherwise it tends to spread a layer of cool air close to the floor while leaving warmer air above, where it counts most.

When installing through-the-wall unit, be sure to caulk seam around unit to prevent inside-outside air leaks.

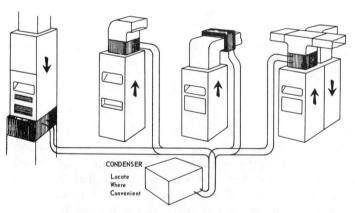

CONDENSER
Locate
Where
Convenient

TYPICAL REMOTE SYSTEMS WITH COOLING COILS

How cooling units are inserted in common types of furnaces (from left): highboy, highboy, extended plenum, lowboy. Condenser and compressor unit may be located at most convenient outdoor point. The nearer to the furnace the better. Refrigerant lines lead from condenser to cooling coil.

Here furnace blower is used to drive cooled air through system. When damper from furnace to plenum is closed, as in summer, blower air is forced through duct to blower coil, and on through ducts. When damper from blower (cooling) coil is shut, as in winter, and furnace coil is open, blower air is driven past heat exchanger in usual manner, and on through ducts. When one damper is closed, other damper must be opened.

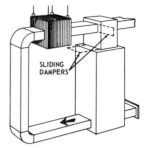

TYPICAL BLOWER REMOTE SYSTEM

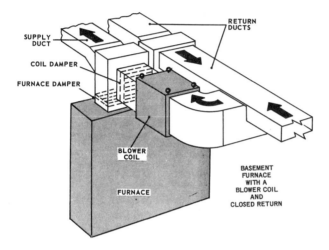

Independent blower coil uses its own blower, doesn't depend on furnace blower, and can be cut in or out of duct system by dampers. In winter, damper from cooling coil is closed, damper from furnace is opened, and cooling blower coil is shut off. Only warm air circulates through ducts. In summer, damper from furnace is closed, damper from blower coil is opened, and furnace is shut off. Only cooled air circulates through ducts.

CENTRAL AIR CONDITIONING takes several forms. If you already have forced warm-air heating, you may have little or no additional duct-work to do, though some modification of the plenum will be required to house a cooling coil. As shown in the drawings, the cooling coil may be installed in an extension of the plenum so that the regular heating-system blower drives air through it for summer cooling. In winter the warm air also passes through the cooling coil, but as the cooling coil is not operating it does not interfere with heating. In some cases, however, it may be necessary to increase the power of the blower motor in order to maintain adequate warm-air flow through the added resistance of the cooling coil.

The "blower coil" can be used when for any reason it is not practical to use the furnace blower for air conditioning as well as heating. The blower coil is mounted in its own plenum along with its own blower and connected to the plenum or main starting duct of the heating system. As

179

shown, a damper (usually a sliding type) shuts it off from the distribution system in winter, when the heating system is in operation. This arrangement has no retarding effect on the flow of heating air, as heating air does not pass through the cooling coil.

The split system is used in central air conditioning because it is seldom practical to mount any type of single-unit air conditioner in a central system. The split system is simply a combination of components that places the cooling coil inside the house (at the start of the duct system) and the condenser and compresser outside—with copper tubing connecting the two parts. This removes most of the possible noise or vibration from the house and places the condenser and compresser where they have an ample supply of cooling outside air.

Because of the difficulties involved in "charging" this type of system with refrigerant in the past, most installations were made by professionals. Today, however, installing these systems is a relatively simple do-it-yourself job because of the perfection of "pre-charged" components and a special type of connection fitting.

Typically, both the cooling coil and the condenser are factory-charged with refrigerant and sealed. The connecting tubing, too, is charged and sealed. When the tubing is connected to the components, however, a cutter inside the fitting breaks the seal between the tubing and the component to which it is being connected, but the refrigerant cannot escape because the other end of the tubing is still sealed. Not until all connections have been made can the refrigerant flow through the system. And at that point, none can escape.

In its most common form, the split system consists of a cooling coil in the starting ductwork and a condenser and compressor located on the ground outside. The outside location should be such as to provide the shortest feasible run of tubing between components. Variations on this system are available in a wide range of models with different capacities. For example, roof-mounted units may be used to provide short duct runs from the cooling coil and blower in a house that does not have a warm-air heating system. This brings the cool air into the rooms at ceiling level, directing it downward and usually providing a mixing action. Heated air is usually removed at ceiling level, too, at a location not likely to draw off any of the entering cooled air.

The most suitable system depends on your house and your own personal feelings about air conditioning. Even an undersized air conditioner contributes to your comfort, not only by its cooling effect but by its dehumidifying action. (You feel cooler when the air humidity is lowered—even if the temperature is not.) But it is not wise to install an inadequate unit, as the correct capacity can usually be had for a relatively small increase in price.

If you are likely to be physically active in your home, as in playing table tennis or turning out major workshop projects, or if you regularly

Split-system units (inside-outside) by Carrier, may be mounted with one unit a considerable distance from the other, interconnected by two refrigerant lines.

Same unit can be bolted together, connected in close-coupled fashion, and used as a through-the-wall unit. It provides 6,000 b.t.u.'s of cooling capacity, enough for large home, town house, or offices.

have a large number of people in the house, you are likely to need a fairly large air-conditioning unit because you will require a lower temperature. You may very well want a constant temperature around 72 or 75 degrees. If, on the other hand, you spend most of your time relaxing you'll probably feel comfortable when the indoor temperature is higher—but below the outside temperature—because your cooling system in most cases will reduce humidity and make you feel cooler than your are. In the final analysis, your air-conditioning unit's capacity depends on what you consider comfortable temperature. To decide what you like, take a thermometer with you when you go to a restaurant or theater, or someone else's air-conditioned home. This won't tell you anything about the humidity, but it will give you a fair guide to what you consider comfortable air cooling.

Improving the cooling situation may be worthwhile in some homes before planning air conditioning. For example, if you have an uninsulated house, old or new, with a dark roof, or if you have big picture windows with a single layer of glass, the effective air-conditioning system may cost a lot to install and to operate. By adding insulation to the area you want to cool, or by using a paint made for the purpose to lighten the color of your roof, or by replacing the single layer of glass in your picture window with a double layer you can reduce not only your initial expenditure but your operating costs as well.

Index